CAKE FORMATION IN PARTICULATE SYSTEMS

Edward J. Griffith

Edward J. Griffith
Detergents and Phosphates Division
Monsanto Chemical Company
800 N. Lindbergh Blvd.
St. Louis, MO 63167

Library of Congress Cataloging-in-Publication Data

Griffith, Edward J.
  Cake formation in particulate systems / by Edward J. Griffith.
    p.   .cm.
    Includes bibliographical references and index.
    ISBN 0-89573-748-5
    1. Particles. 2. Agglomeration. I. Title.
TP156.P3G75 1991
660'.293—dc20   91-7709
                                                                                              CIP

British Library Cataloguing in Publication Data

Griffith, Edward J.
  Cake formation : the applied chemistry of cake formation
  in particulate systems.
  1. Particulate solids
  I. Title
  660.2842

ISBN 3-527-27844-3

ISBN 0-89573-748-5

© 1991 VCH Publishers, Inc.

This work is subject to copyright.

All rights are reserved, whether the whole or part of the material is concerned, specifically those of translation, reprinting, re-use of illustrations, broadcasting, reproduction by photocopying machine or similar means, and storage in data banks.

Registered names, trademarks, etc. used in this book, even when not specifically marked as such, are not to be considered unprotected by law.

Printed in the United States of America.

ISBN 0-89573-748-5 VCH Publishers
ISBN 3-527-27844-3 VCH Verlagsgesellschaft

Printing History:
10 9 8 7 6 5 4 3 2 1

Published jointly by:

VCH Publishers, Inc.	VCH Verlagsgesellschaft mbH	VCH Publishers
220 East 23rd Street	P.O. Box 10 11 61	(UK) Ltd.
Suite 909	D-6940 Weinheim	8 Wellington Court
New York, NY 10010	Federal Republic of Germany	Cambridge CB1 1HW
		United Kingdom

DEDICATED

TO

Professor Harold E. Wilcox

whose teachings are found throughout this book

CONTENTS

PREFACE

CHAPTER ONE
Introduction
Concretus Chemistry 2
The Cost of Caking in Time and Money 6
Physical Forms 9
What Is the Applied Science of Caking? 11
How To Use This Book 11
References 14

CHAPTER TWO
The Classes of Cake Formation
The Four Classes of Particulate Caking 17
   Mechanical Caking 17
   Plastic-Flow Caking 19
   Chemical Caking 19
   Electrical Caking 20
Variables in Cake Formation 21
   Composition 21
Subclasses 23
   Basic Components 23
   Acidic Components 23
   Volatile Components 23
   Low Melting Point Components 24
   Unstable Components 24
   Weak Solid Structures 24
   Heat Capacity and Heat Conductivity 25
   Seasonal Influences 26
   The Kinetics of Cake Formation 26
Compression 27
The Flow of Solids 29
Product Types 30
Summary 31
References 31

CHAPTER THREE
The Chemistry of Cake Formation
Chemical and Physical Changes 33
Water 39

Adsorption 42
  Case 1 45
  Case 2 45
  Case 3 46
  Case 4 46
  Case 5 47
  Case 6 47
  Case 7 48
  Case 8 48
  Case 9 (The solution surface) 48
Colloid Science 49
Hydrates 51
Amorphous Solids 58
Physical Property Measurements 62
Solubility 62
General Rules 63
The Phase Rule 65
Heats of Solution 66
Mixed Systems 69
Grinding 70
Formulations 71
Detergents 73
The Flow of Solids 77
Summary 77
References 79

CHAPTER FOUR
Phase Behavior and Cake Formation
Phase Transitions 82
  Polymorphic Changes 84
Phase Diagrams 91
Degrees of Freedom 91
  Two Solid Components Diagrams 92
  Three-Component Three-Dimensional Diagrams 102
  Three-Component Aqueous Diagrams 103
Summary 112
References 113

CHAPTER FIVE
Electrically Induced Cake Formation
Electrical Crystals 114
Electrical Caking 115
Mechanisms for Piezoelectric Caking 116
Pyroelectric Crystals 121
Seed Crystals 121
Ferroelectric Crystals 123

Contents / vii

Static Electricity 126
   Charges on Crystals 127
   Positive and Negative Charges 129
Flow Conditioners 132
   Charges on Liquids 133
Summary 135
References 135

CHAPTER SIX
Laboratory Techniques and Test Procedures
For Managers 137
Laboratory Personnel 138
Project Duration 138
Laboratory Versus Plant 139
Reports 140
The White Powder Syndrome 141
Environmental and Health Issues 141
Types of Laboratory Tests 142
Sample Selection 143
Statistics and Experimental Design 144
Experimental Design And Caking 145
Water Absorption 147
Phase Chemistry 149
Solubility 150
   Three Component Aqueous Phase Diagrams 152
   Two Component Melt Phase Diagrams 154
Weight Loss on Ignition 160
X-ray Analyses 161
Microscopic Studies 161
   Refractive Index 162
   Television Equipment 163
Electrical Properties of Crystals 163
   Piezoelectric Test 165
   Pyroelectric Test 166
   Ferroelectric Test 167
Summary 168
References 169

CHAPTER SEVEN
Flow Schemes to Classify Caked Solids
Types of Schemes 170
Scheme of Attack 171
CAKING FLOW SCHEME NOTES 179
   Notes for Scheme No. 1 179
   Notes for Scheme No. 2 179
   Notes for Scheme No. 3 180

Notes for Scheme No. 4   *180*
Notes for Scheme No. 5   *180*
Notes for Scheme No. 6   *181*
Notes for Scheme No. 7   *182*
Summary   *184*

CHAPTER EIGHT
Typical Solutions to Caking Problems
Commercial Problems   *185*
Ammonium Nitrate   *187*
Fertilizers   *191*
Ammonium Phosphate   *192*
Ammonium Sulfate   *192*
Urea   *193*
Phosphate Fertilizers   *195*
Detergents   *197*
Foods   *198*
Sugar   *199*
Salt   *200*
Swimming Pool Chemicals   *201*
Water Treatment Chemicals   *203*
Metals Treating   *203*
Summary   *205*
References   *205*

CHAPTER NINE
Induced Cake Formation
Scope   *208*
Agglomeration   *209*
Prilling   *213*
Spray Drying   *215*
Microencapsulation   *216*
Tableting   *217*
Casting   *218*
Compaction   *218*
Summary   *220*
References   *221*

CHAPTER TEN
Overview and Outlook
Future Improvements   *228*
Needed Laboratory Work   *228*

PREFACE

THE APPLIED CHEMISTRY OF CAKE FORMATION IN PARTICULATE SYSTEMS

A Generalized Description

Cake formation in solid products is so common that hardly anyone is immune to the problems that cake formation can cause. It may be as simple as salt that will not flow from a salt shaker to the return of an entire warehouse of finished goods to the manufacturer because his product has become one gigantic lump.

Solids cake and form lumps for a wide variety of reasons, but most caking phenomena can be classified under one of four major types. When the type of cake has been identified, work to eliminate the problem can be initiated in a predictable, organized approach. Once the cause for cake formation is understood an assessment of the probability of success can be made with greater confidence. Not all caking problems can be eliminated under all conditions. The classical principles underlying cake formation are taught in this book in a simplified manner, requiring a very limited knowledge of mathematics, chemistry, and physics to understand the basic concepts. The goal is to give the reader a knowledge of the basic science governing cake formation, while offering a standardized procedure to determine the critical factors involved in any caking problem. This book was written for applied scientists with pragmatic, profit-stealing caking problems. An attempt has been made to include concepts with direct applications, while avoiding the temptation to treat each subject in detail. This means that a number of important subjects must be handled incompletely, but to cover them in the manner they deserve would be much more ambitious than the intended scope of this book. This constitutes the author's apology for this deficiency and no additional mention will be made of this point.

Edward J. Griffith

CHAPTER ONE

Introduction

For any product to survive in an environment of global competition, a quality product at a competitive price is imperative. Any industry producing powdered solids, whether they are foods, detergents, coal, ceramics, cements, explosives, dyes, pigments, fertilizers, or chemicals, cannot consider their products as Quality Products if these products arrive at a customer's home, plant, or work site caked and lumped to the degree that the product is not ready for immediate use. When schedules are interrupted or the product must be disposed of rather than used, a customer cannot be pleased. Disposal can be a very expensive if a large quantity of even a safe substance must be scrapped. It may be less expensive to send the product back to the manufacturer to be reworked. In all cases the aggravation caused by products that do not perform as expected is enough to kill a product which might otherwise perform in an outstanding manner.

Few quality properties of a powdered product are any more obvious than lumps in a package or box that is expected to be free flowing. It is almost universally true that lumped materials will be considered poor quality materials and if the customer can find a manufacturer that can supply the same material in a more desirable physical form, he will turn to the other supplier.

The goal throughout this book is to assist the manufacturer of powdered products to produce a quality product at a reasonable cost. The work will be directed toward the chemistry and physics of caking, but packaging and transportation will also be considered. Ideally, a universal solution to all caking and lumping problems is desired but it is unrealistic to believe that this can be accomplished within the restraints imposed on most products. There is small doubt that all caking and lumping problems can be solved if cost is not a factor.

CONCRETUS CHEMISTRY

Most of the history of Earth had already transpired billions of years before mankind made an appearance. Igneous and sedimentary rocks were plentiful and were at least observed, if not understood, by early man. Sedimentary rock formation represents a form of caking where individual particles have cemented together to form a larger coherent body. The outcroppings of these rocks were not unlike the outcroppings that can be observed today.

It is a reasonable assumption that one of mankind's earlier scientific observations was made when he noticed that some particulate matter in his environment would form cakes, especially if they became wet. He probably also noticed that other substances would not form cakes regardless of what he did to them. There was something strangely different about the two kinds of substances. One kind retained his footprint when it dried, while the other retained no imprinted form at all. One kind of substance became hard like stone, while the other kind of substance may have become a dust, blown away by the wind. Modern man is somewhat more advanced than his ancestors but today's knowledge is far from complete and often reasoning is more by association than by fundamental understanding.

The words caking and lumping are poorly descriptive and scientifically unacceptable. Both words have multiple meanings and are in common usage to depict concepts as unrelated as cooking cakes to slang instructions to ignore conditions or events; "Like it or lump it." Even when the subject is clearly understood, it is difficult to forge a consensus as to where boundary conditions should be imposed on definitions. In the context of this book caking or lumping shall be used interchangeably and for the most part shall connote an undesirable condition. A word is borrowed from Latin, *concrescere*, to describe the sticking together of particles and *concretus chemistry* to describe the science, which shall be employed to understand and hopefully prevent the undesired changes from powders to lumps.

This leads to the need for a definition of what is chosen to be called caking. In some treatments, which attempt to be very scientifically rigorous, much ado is made about the differences between adhesion, cohesion, agglomeration, sticking, auto adhesion, flocks, lumps and on and on. It is the desire to be both helpful and informative to the reader but to treat caking and lumping with no great scientific piety. There seldom is elation when a product cakes and from this perspective it is usually a quality of negative value.

Caking is a manifestation of the influence of gravity on solids. In a weightless environment solid particles that were neither sticky nor mechanically entangled should have little tendency to form cakes. Begin with a simple definition of caking.

When two or more macroparticles, each capable of independent translational modes, contact and interact to form an assemblage in which the particles are incapable of independent translations, the particles are defined as caked.

To be completely rigorous the definition should contain several restraining statements with respect to force fields, shearing forces and yield values. As stated it is tacitly implied that the force acting on the system is gravitational and that the system suffers no impacts sufficiently great to rupture the assemblage formed by the macroparticles. In this respect it is desirable that powders and granules emulate liquids as much as is possible, particularly when it is desirable that the particles be placed in motion. It will usually make little difference whether or not a substance is caked if there is no intention of ever having it move or flow.

Controlled caking is often intentionally utilized in processes where it is desired to agglomerate, sinter, prill, pill, nodulize, glue, cement, and so forth, but problems occur when products that should flow freely will not flow from the container in which they are confined, without the application of some external force other than gravity.

Phase transitions are usually involved in the caking of crystalline solids and water is also likely to contribute. But there are numerous examples where caking occurs in completely anhydrous systems. The phase transition may be as simple as the formation of a hydrated salt from an anhydrous salt, if indeed this is simple, or as complex as the dehydration of a hydrated salt or the double melting point of a racemic mixture.

The solubility of a salt, the energy of hydration, the heat of crystallization and hygroscopicity can be important factors. The electrical properties of a crystal can also become involved. Both bridging in conveyors and electrostatic lumping in bottles, cars, and bags can cause problems. Some systems that cake badly are not caked at all in the normal sense but are held by polar interactions, magnetic or electrical. This is to say that there is no physical bridge between particles. Bag set and bottle set can lead to bridged caking, however. Bag set and bottle set will be defined by example. If dense dry ammonium nitrate prills are allowed to stand in a well-sealed bottle for an hour or more, a cake will form in the bottle. The cake can be broken up by shaking the bottle until the prills are free flowing. If the bottle is allowed to sit quiescently, the lump will reform. The cycle can be repeated endlessly with no degradation of the prills. This behavior is caused by an electrostatic charge on the crystals or prills.

Amorphous solids can have yield values low enough to allow the substances to flow under gravitational force. This kind of caking occurs when granules of tars or waxes become warm enough to flow together. A commonly encounter manifestation of this type of caking occurs when gelatin capsules, of the type used to contain vitamins, stick together from one day to the next. Usually the cake will disintegrate when the bottle is bumped sharply, but one capsule stuck to the bottom of the bottle may, at times, be very difficult to dislodge.

Mechanical caking can range from systems as simple as coat hangers or fish hooks to the entanglement of cotton in bales. Some very interesting studies that deal with the formation of knots have been published [1]. It has been shown that there is a general mathematical polynomial to describe knots. Although detailed descriptions of entanglements are not possible at this time, much progress has been made in relating factors as bulk density and aspect ratio [2]. Aspect ratio is the length of a particle divided by its diameter. It is not unlikely that at some future date it will be possible to describe mechanical caking mathematically. Velcro is a useful two-dimensional example of mechanical caking.

Because modern society continues to be plagued with caking problems, many scientific papers have been published which deal with specific problems of caking. Some excellent treatises have been written which offer detailed coverage of the science of particle interactions and the physics of long range and short range forces. *Adhesion of Dust and Powder* by Anatolii Zimon is a classic [3]. An understanding of these works is imperative for an in-depth knowledge of the interaction of particles and the media in which they are contained, but the technicians faced with their own unique caking problems may have no place to turn for help. It is a prime objective of this book to offer a starting place for those faced with caking problems, but who have no real desire to master statistical mechanics in order to eliminate a caking problem in a box of soap powder, a sack of sugar or a bag of fertilizer. Some established references on the mechanics of powders have been consulted both for definitions and concepts. *Principles of Powder Mechanics* by Brown and Richards presents many fundamentals of powder technology in an practical and understandable form [4]. The approach presented here is less rigorous but depends heavily on the background information of the established literature.

The science of caking will be approached from a less mathematical basis than is found in the physical treatments. This is to say the subject will be approached more from the view point of the chemist rather than from the view point of the chemical physicist. Alexander Findlay notes in the preface to his classic masterpiece, *The Phase Rule and Its Applications*, "Although we are indebted to the late Professor Willard Gibbs for the first enunciation of the Phase Rule, it was not till 1887 that its practical applicability to the study of Chemical Equilibria was made apparent. In that year Roozenboom disclosed the great generalization, which for upwards to ten years had remained hidden and unknown save to a very few, by stripping from it the garb of abstract Mathematics in which it had been clothed by its first discover" [5]. Every effort will be made to avoid this pitfall. Solubility, vapor pressure, hydration, phase transitions, thermal history, crystal properties and electrical behavior will be covered in detail and though scientific in their content they are more easily visualized than concepts presented as abstract mathematical equations. This is in no way intended to imply that a physical approach is not vital. It is! The reader is encouraged to read some of the in-depth treatments listed with the references.

Many industries have been plagued with caking problems for decades and one sure criterion of a quality product is whether or not it cakes under normal use conditions. If a product cakes, it is a reasonably good indication that the formulators did not have a clear grasp of the physical chemistry necessary to control the physical properties of their products. To be sure, environmental issues have made superior powdered products almost impossible to manufacture but satisfactory products can be manufactured, particularly when there is no longer a truly superior bench mark by which the customer can judge performance [6]. It can be stated with reasonable certainty that no other product on a grocer's shelves has had more scientific research expended upon it than the detergents. Fortunately, the formulated products have behaved well enough in their boxes that only a small fraction of the research has been spent on caking problems.

The institutional and industrial segments of the detergent industry, are intentionally casting powdered detergents into blocks. This yields a product that is easily dispersed in automatic equipment and avoids the problems associated with caking. The detergents used in the industrial and institutional markets are very similar to the detergents used in households in years gone by.

The problems associated with lumping and caking are so universal, and in some industries so commonplace, that they are accepted as a part of the business. Some products are cast in drums when manufactured and giant "can openers" are used to remove the drums from the products. As a result of these circumstances, it is difficult to obtain records of the cost of caking. In years passed, the fertilizer industry treated caking as a part of farming. A farmer expected to beat a bag of fertilizer with a sledge hammer before dispersing it on his soil. Parts of the industry intentionally caked the products in bins during manufacture and depended on milling the product before bagging it [7]. This allowed hot products to cool through transitions or damp products to hydrate fully before the product was milled. The storage of a product before milling was usually referred to as aging the product. Although aging was usually successful, it was expensive because a backlog of product was held in inventory for hours or days before it could be milled.

Milling could include blasting the solid blocks with dynamite in order to break it into particles that could be milled with conventional milling equipment. In some cases the fertilizer industry has led the way to liquids rather than solids and in the case of anhydrous ammonia even gases have been used. It is often much easier to distribute a liquid than it is a solid but care must be exercised by the consumer to be certain that the cost performance of liquids is at least as good as the cost of performance of solid fertilizers. In this case performance is usually measured by the yield per acre and grade of the harvest.

For many years ammonium nitrate has been a part of both the fertilizer and the munitions and blasting industries. Ammonium nitrate caking has been an example of most of the possible problems a granular or prilled product can cause when caking occurs. Sometimes when ammonium nitrate was a part of a cake, disastrous results were obtained when the cake was blasted in an effort to reduce the cake to lumps that could be milled. The caking of ammonium nitrate will be given considerably more attention in the chapter dealing with typical examples.

THE COST OF CAKING IN TIME AND MONEY

A part of the difficulty in estimating the time and money that is lost each year as a result of caked products is that the indirect expense is probably as great as the product loss itself. Production schedules are missed. Products are returned to the manufacturer to be either reworked or discarded with transportation cost and paper work on both ends of the transaction. As previously mentioned, it is often difficult to find economical means of disposing of a large quantity of a commodity and the substance may have to be reworked, regardless of cost. In this way a substance that would otherwise be regarded as waste can be spread over an entire nation or the globe to perform a useful service, while it is disposed of in a more acceptable manner. In this way we may literally eat our own waste as an acceptable way of disposing of it but at a very high price to the public who ultimately pays this bill.

When all factors are considered, the cost of unproductive cake products is probably in excess of one billion dollars per year in the United States alone. At the time of this writing there was a trainload of finished powder that was caked in its railroad shipping cars. It had been shipped from the northeastern United States to Mexico and back with no customer desiring a product that cannot be unloaded by the more or less standard methods. More time and money will be spent on the inferior product than a new product meeting specifications should cost. Only one product in the plant of a major producer of food-grade products was estimated to have cost the company over one million dollars last year, as a result of caking problems. Bauder reports that in a plant producing a powdered gravy food product that twenty-five percent of the downtime of the plant's packaging line was attributable to caked product [8].

Help was requested from the Stanford Research Institute and several agencies of the United States Government in an effort to arrive at a number that was something more than a guess as to how much it cost the people of the United States each year because of product caking. After a considerable effort it was concluded that an authentic value could not be calculated. Rather than leave the question completely voided, a value will be calculated based on some assumptions that are believed to be conservative.

The first assumption to be used in estimating the cost of caked products in both direct and hidden cost is, 0.5% of the cost of all finished powders and granules is directly related to caking problems. For typical systems the following values can be calculated.

The data in the following table are presented to point out the magnitude of the problem and are not intended to be anything more

TABLE 1

Industry	Estimated Value of Goods* (in billions of dollars)	Cost of Caked Goods (in millions of dollars)
Drugs, soaps, etc.	53.9	269
Fertilizer	8.9	44
Inorganic Chemicals	68.7	344
Stone, clay, glass	55.0	275
Sugar and salt	4.2	21
Total value 1985	190.7 billion	953 million

*Estimates based on data from Statistical Abstracts of The United States 1986

than estimates. The difficulty in obtaining reliable data does bring attention to the need for the collection of data of this type by some reliable organization. Perhaps the United States Department of Commerce would be an appropriate organization. With the collection of data, it would become evident that either more work should be committed to understanding the caking phenomenon or that the problem does not consume a sufficient portion of the national resources to make it worthy of the research required to solve the problems of industry and the citizens of the world.

Large industries have been established to manufacture equipment exclusively designed to handle powdered or granular solids. Rail cars and trucks are but the beginning of the equipment list. Belt drives, shovels, air conveyors, or screw conveyors are in common use, as well as a variety of shakers and pneumatic or electrical hammers. Some very large equipment is capable of emptying a railroad freight car or a truck by turning the car on its side and dumping its contents into a suitable bin. Depending upon the type of storage, if any, to which a product will be subjected, it may be placed in a silo or even bagged. If bagged and placed in a warehouse, any number of circumstances can influence the bag during storage. The season can play a very important role in the history of the product. In the spring and summer months, the temperature of a stored product is likely to experience wide swings from day to night, particularly if the product is stored on the side of a warehouse that receives the afternoon sunshine. In the spring and summer months the partial pressure of water vapor in the air surrounding the product is likely to be much higher. Products that behave very well during winter months may give many problems as the conditions change in the spring. Problems that occur as a result of static electricity may worsen in cold dry conditions. If the product contains any free water or hydrates which decompose at relatively low temperatures, water will migrate from the hot side of a container toward the cooler parts of the container where it is likely to hydrate lesser hydrated salts or anhydrous crystals. In either event a phase transition has occurred both when the hydrate decomposed and when the new hydrate formed.

If a product has been stored in bags, other variables become active. The surface area of the product is usually increased many times over the surface-to-volume ratio of bulked materials. This means that heat transfer to the surroundings will change. The tendency of water to migrate will change. The area to absorb water from the atmosphere will increase if the bags are not well sealed against moisture. Most bagged products are stored in stacks on movable platforms (dollies). The location of a bag in a stack must be considered. Bags placed at the bottom of a stack may be subjected to tons of force, while the bags at the top of the stack are likely to be subjected to the greatest temperature or water vapor variations. Not only is the type of bag very important, but also the color and materials is worthy of attention. These factors are of importance, particularly to a product that is hygroscopic. In fact, the way the bag is sealed may become an important consideration. Heat sealed plastic films usually are more resistant to vapor penetration than are sowed seals. In some cases it is impractical to attempt to prevent water vapor from penetrating a container bag. Very hygroscopic salts will cause water vapor to penetrate even a well-sealed bag. Water vapor may permeate bags that are impervious to liquid water.

All too often insufficient attention is given to the physical chemistry of a product when selecting a site for a plant to manufacture a product. The classic examples are the ammonium nitrate plants that are sometimes built in a swamp. When the wrong location is chosen for a plant, the product suffers for the entire life of the product and factors such as freight and site cost can be quickly lost to reworked product or expensive refrigeration equipment required to dry the process air. There is a long list of considerations that must be weighed when deciding where to build a new plant. The physical properties considerations, too far down the list, particularly if the product is very soluble in water, can destroy an otherwise outstanding product.

PHYSICAL FORMS

Many single chemical products are sold in a variety of physical shapes and bulk densities, depending on the requirements of the customers. The chemical, sodium tripolyphosphate, is a classic example. There are three forms of the product that depend upon the phase chemistry of the salt. There is the high temperature phase, Form I, the low temperature phase, Form II, and the hexahydrate resulting from the hydration of either Form I or Form II. The chemical properties of the three forms are so dramatically different as to cause it to be difficult to believe that all three forms when dissolved in water yield identical solution properties.

The sodium tripolyphosphate is sold as high or low Form I content powders; high, medium, or low density granular; partially or totally hydrated salts or combinations of most of the above. In the trade the grades are usually referred to as high and low TR where TR means temperature rise during hydration and is related to the quantity of Form I sodium tripolyphosphate contained in the product. In any case the customer cannot tolerate a caked product or one that unloads extremely slowly. At one plant where accurate records were maintained, over a three-month interval, rail cars required between six and *ninety* hours to unload. The man-hours spent in unloading the caked cars was in excess of the actual hours, adding an unacceptable price to the cost of goods.

In all systems the higher temperature phase of a product will have a greater release of energy when hydrated or dissolved. It is the extra energy added to the salt when it was converted from the low temperature phase to the higher temperature phase. Also in most equilibrium, hydrated salt systems, if the salt can form more than one hydrate, the hydrate containing the greater number of water molecules will be the lower temperature hydrate.

Obviously, high, medium, and low density granular, high and low Form I powders, and hydrated and partially hydrated powers cannot all have the same bulk handling properties. But if properly manufactured and delivered they can all be supplied as free-flowing solids.

It is possible to completely destroy a correctly manufactured product by shipping it in a car or truck that leaks water or snow into the product. The car may even contain water from the last shipment. Too often, the untrained operator loading the cars is not aware of the devastation that can be caused by rainwater leaking through a defective seal. If this is the weak link in a process, all of the sophisticated science and money that may be thrown at the problem will not be as effective as teaching an operator what action should be taken when defective shipping equipment has been supplied. In many plants an operator does not have the authority to refuse a defective car or truck.

Little can be done in a laboratory to overcome poor plant practices. In the very honest efforts to meet production schedules, it is not uncommon for plant personnel to cut corners to meet the schedule. Many salts have physical crystalline phase transitions when cooled from a high temperature to ambient temperature. If a product is loaded into a car at a temperature above its transition temperature, it is certain to cake. Because most salts are more soluble at higher temperature, any water contained in the product will cause caking when the product is loaded hot. Most processes dealing with heated solids must be equipped with adequate coolers if quality products are to be supplied to customers.

Operators attempting to perform their jobs well may load materials that are too hot in an effort to meet production schedules. They can make disastrous decisions if not properly instructed and trained.

A guard at a research laboratory told of his experiences as an operator in a chemical plant during World War II. He related how poorly many chemical engineers understood their processes. While operating a process he and the other operators learned that if they ran the reaction at only ten degrees above the specification temperature, the reaction was completed in only half the time. Consequently, he and the other operators could play cards for the second half of the shift. The operators had been preparing hundreds of tons of *gun cotton!* each shift. In effect, he and his friends had removed all of the safety factor the engineers had built into the process. Free-flowing solids begin with trained operators. This should be the first order of business for anyone faced with a caking problem.

WHAT IS THE APPLIED SCIENCE OF CAKING?

During the latter part of the twentieth century it has become more and more difficult to describe what is science and what is not. Many studies are classed as science that only a few years ago would have not have been dignified with a proper noun. When an attempt is made to subclass a science as either pure or applied, the task becomes even more difficult.

There is no specific body of knowledge that can correctly label the science of caking, either pure or applied. It is also unlikely that a division of any established scientific society shall ever be named the Division Of Caking Science. Too few scientists are interested in the subject and too little scientific literature is devoted to the topic. The work that has been done is almost exclusively the product of industry and much of the work has been directed toward one particular product with little or no attention directed to other products. Much of the academic research that is published is a part of the beautiful art forms of science. At best, lumping and caking can hardly be considered a scientific art form.

The mechanical tools of this work will also be as simple as the theory. Most of the instruments employed are common in most laboratories. The concept is carried throughout this work that research is a thought process and bench work is done to prove or disprove the thought process while obtaining the quantitative data to support or reject the concepts. The solving of any caking problem is certainly in the domain of research; whether or not it is a science depends upon the point of view.

HOW TO USE THIS BOOK

As background to the use of this book, it should be helpful to know why the book was written. In the early 1950's one of the leading manufacturers of detergents was having caking difficulties with its leading detergent. The company approached the Monsanto Company, a raw materials supplier, to determine whether or not they had a scientist who felt that he could determine the quantity and location of water in a complex mixture of crystalline and amorphous, inorganic and organic compounds that had become known as a synthetic detergent. A Chevenard thermobalance was being used as a research tool to analyzing water in mixed hydrates. The approach worked well in the detergent. The problem was soon solved.

Some months later Monsanto acquired Lion Oil Company and its ammonium nitrate business. Caking problems in ammonium nitrate are legendary. They desired to manufacture a new form of ammonium nitrate that would not cake as a result of phase transitions. Some stored samples of ammonium nitrate more than twenty years old show no signs of caking.

This is primarily a reference book. It is intended to be readable without the reader being forced to master difficult theory to understand the contents. If one has a long-term caking problem that requires a solution, it is recommended that the entire book be read rapidly while noting sections that seem to be related to your particular problem. The parts that refer to your problem should be read more carefully and if more theory is required, it will probably be found in the references. Much of the published work dealing with particulate behavior has been divided into two types. One type has been highly theoretical, the physics of solid interactions, while the other has been test- and equipment-oriented with little or no theory. An attempt will be made to work between the two extremes, mixing some theory with a cookbook laboratory manual approach.

Each of the chapters has been written in a stand-alone mode as much as possible. The advantages are that any section can be used without the reader being forced to read all other chapters. The disadvantage is that a certain amount of redundancy is necessary when a concept is addressed in more than one place. Two types of chapters have been written. Most of the book is general information and background. Two chapters are intended to be hands on, cookbook directions. These are Chapter VI, "Laboratory Techniques" and Chapter VII, "Flow Schemes." When the necessary text has been understood, the flow schemes should be utilized with the aid of the chapter on laboratory techniques. The flow schemes are similar to computer flow charts or flow diagrams used in qualitative analysis schemes. These schemes begin with the proper methods of obtaining samples for study as utilized by professional analytical chemists. Then the samples are subjected to a battery of tests that are organized to identify the kinds of caking exhibited by the sample. As one progresses through the schemes, one caking mechanism after another will be eliminated until the likely mechanism active in your sample is identified. It is recommended that the sample be passed through all seven schemes even if one mechanism of caking has been positively identified. In some cases more than one kind of caking may be active.

Usually the process of testing the sample will suggest a variety of potential solutions to the caking problem. If a solution does not come forward as a result of the test, it is then recommended that the chapter dealing with typical solutions be consulted. It is best to begin by choosing solutions that have chemistry similar to your sample. It is possible that the particular problem is unique and none of the known solutions are similar enough to be of much help. If this is the case, the only choice is to begin new research aimed at solving the particular problem.

While solving caking problems, it is better to forget whether or not a solution to the problem is practical at first. After successful solutions have been demonstrated, even overly expensive or impractical solutions can be used as pilot demonstrations. In the early experimental phase of any problem solving, extrapolation to the extremes, using a broad brush approach, is recommended. The more familiar the investigators become with how their systems behave under extreme stresses, the more likely the investigators are to solve their problems. For example, if an investigator has reason to believe that an impurity in any component of a product could be causing the problem, an effort to obtain an ultra pure sample may be well worth the expense. If the ultra pure sample is well behaved, then the known impurities can be added one by one until the culprit is exposed. For many years this technique has been remarkably successful in troubleshooting electronic equipment and the technique is proving to be equally successful in isolating problems in modern computers [9]. Of course, this technique performs best when single compounds are the product, but it can be applied to formulations provided interactions in the formulations are first understood. This may or may not be a simple task. Interactions shall be discussed under the section, "Phase Chemistry." Although the interactions may be isolated by some type of statistically designed experiment, the interactions involved are more physical than mathematical. The statistical approach will be discussed in more detail at several places, when it is appropriate.

In discussing phase chemistry in this book the Phase Rule will be used freely but will not be stressed, not because it has no use, but because the Phase Rule seems to confuse more readers than it helps. Phase chemistry existed long before there was a Phase Rule [5]. The Phase Rule placed phase chemistry on a firmer thermodynamic footing and dispelled some ill-conceived notions about phase equilibria, but phase chemistry, as presented, can be readily understood without referring to the Phase Rule, except in an oblique way. Its use will be demonstrated from time to time when it will be helpful in understanding the behavior of a system. The reader is encouraged to become conversant with the Phase Rule if it has not already been studied. It lends much to the understanding of the interaction of solid phases.

The references have been chosen from books the author has often found to be helpful. This does not mean that the current literature will be neglected, but the modern computer search allows one to obtain hundreds of references in a short time. In preparing for the actual writing of the book, the author obtained about two thousand references that were related to the subject in some way and the process required less than one day's work.

REFERENCES

1. Peterson, I., *Science News 133*, 329 (1988).
2. Katz, H.S., and Milewski, J.V., ed., *Handbook of Fillers and Reinforcements for Plastics*. Van Nostrand Reinhold Company, New York (1978).
3. Zimon, A.D., *Adhesion of Dust and Powders*. Consultants Bureau, New York and London (1980) (Trans. by Robert K. Johnston).
4. Brown, R.L., and Richards, J.C., *Principles Of Powder Mechanics*, Pergamon Press, New York, London (1970).
5. Findlay, A., *The Phase Rule and its Applications*, Dover Publications, New York, N.Y. (1st ed. 1904) (1938).
6. Odioso, R.C., *Happi, October,* p. 122 (1988).
7. Bookey, J.B., and Raistrick, B., *Chemistry and Technology of Fertilizers*. Edited by Sauchelli, V. Reinhold Publishing Corporation, New York p. 454 (1960).
8. Bauder, U., *Food Engineering International 3,* 23 (1978).
9. Margolis, A., *Troubleshooting And Repairing the New Personal Computers,* TAB Books Inc., Blue Ridge Summit, Pa. p. 216. (1987).

CHAPTER
TWO

The Classes of Cake Formation

Numerous articles dealing with the theories of the caking of solids have been written [10]. Perhaps the chief criticism of these articles is not the quality or accuracy of the information but the scope. Most articles recognize the role of moisture in caking, but too few recognize phase transitions, piezoelectric and pyroelectric behavior, or disproportionation and carbonation reactions as causes. Gross classifications will be made of systems that have caking mechanisms in common. Next, the detailed behavior that causes each system to behave in a unique and identifiable manner will be studied.

It is the object of the classifications to first suggest how a problem area should be approached, which tools should be marshalled to attack the problem, and how similar problems have been solved in the cases where solutions are known. When a product exhibits caking problems, attention will be directed toward changing the behavior of the product rather than changing the bulk solids handling equipment employed for the product. Some references will be made to conveyors, silos, bins, and so forth in the chapter dealing with typical solutions. This is not intended to imply that one approach is any more important than the other, but that a detailed review devoted to industrial equipment is beyond the scope of this book.

Caking will be divided into four major classes in order to give insight into the causes of caking and the classification will suggest the types of approaches that should be considered while attempting to find a cure for a problem. The classes are plastic flow, mechanical, chemical, and electrical. More than one class of caking may be active in a given system. It is possible that an electrical interaction may occur before a system undergoes a dehydration or a physical phase transition and the electrical interaction merely sets the system to enhance bridging between particles.

The plastic-flow and mechanical entanglement problems are obvious to an observer, but the fact that they exist may or may not suggest a solution to the problems they may cause. Chemical caking can become very involved but is probably the most easily solved. Hydrations, dehydrations, chemical reactions, phase transitions, solubility, hygroscopicity, carbonization and similar chemical behavior will be introduced in this chapter. Electrical caking is not an obvious class and requires some special techniques to detect and to eliminate.

Because there are so many different manifestations of caking and lump formation, the first problem to be faced after it has been definitely established that a problem does exist is to identify the class of caking that is occurring. There are many subclasses to each of the major classes but these subclasses usually have some properties in common. All too often, much time and expense is wasted attempting to solve the wrong problem. It is also possible that more than one class of caking is functioning at the same time and the actions confuse the results.

Most forms of serious caking involve some kind of physical bridging between the particles contained in the product. This means that there is a mass transfer across a boundary between two particles that are in the process of caking. It may be that one should consider the boundary moving and the molecules or ions remaining more or less where they were. Mass transfer is probably a characteristic of all caking with the exception of mechanical caking, which is really entanglement. Some varieties of electrical caking may not involve mass transfer but other varieties may well involve some mass transfer at crystal interfaces. The argument will be made that some forms of electrical caking involve the transfer of matter or charge across the boundary of two particles, causing them to unite. Plastic-flow is perhaps the most direct illustration of the concept. Two particles of a wax or tar are in contact. The temperature is so low that the yield point of the solid is not exceeded and the particles remain free flowing. As the temperature of the particles rises, the yield point of the solid is exceeded and the molecules at the interface begin to flow into each other. When the temperature is decreased again, the particles have merged and a mechanical bridge has formed between the particles, if indeed, they can still be recognized as the two original particles.

All forms of bridged caking then require some form of mass transfer across the boundary interface of the particles. Sometimes the systems may melt to cause the flow, as a transition at an interface. At other times a solvent vehicle may assist the molecules or ions to cross the interface, as the formation of a solution, and at other times a sublimation or distillation process may be at work. It will be seen that many of the flow conditioners place a shield across the interface between particles to prevent mass transfer because the particles of the system can no longer contact. The flow conditioners can also counteract surface charge on particles. This book shall be directed toward the identification of the classes of caking and an understanding of what is happening at the interface between particles in hopes that understanding will lead to a practical elimination of the caking problems.

THE FOUR CLASSES OF PARTICULAR CAKING

The four classes of caking are remarkably differentiated and it is seldom difficult to identify a caking problem as belonging to one or more of the four classes. There are, however, many contributing factors that can influence caking and lumping of a solid. These other factors can be considered as subclasses that have contributed to the cake formation. Much of the work before us will be directed toward first determining which class of caking is involved and then isolating the particular properties of the system that are contributing to the problem. Hopefully, the investigator armed with this information can determine the severity of his problem and what, if anything practical, can be done about it.

No one class of caking is more important than any other. The class of caking with which one is currently confronted is usually considered to be the most important. Any caking problem can be very difficult to solve in a practical way. This point will be repeated many times in a number of ways. All caking problems can be solved but there are times when the solution is unacceptable. Cost will usually be the factor that makes a solution unacceptable. No one is likely to refrigerate a fertilizer to prevent it from caking, but if fertilizers could be handled as frozen foods are handled, most caking problems would cease to exist.

Mechanical Caking

Mechanical caking will be considered first, not because mechanical caking problems are easily solved, but because they are very easily recognized. Most people would probably not consider a wad of cotton as caked cotton nor a sheet of tissue as a cake of fibers but, in fact, they are. Everyone is familiar with the entanglement of coat hangers. Fish hooks and fishing line also present similar problems.

The analysis of all of the factors that contribute to "bird nesting" in fibers and paper pulp is involved and difficult. No treatment of the subject has been completely satisfactory, although some very interesting mathematical approaches have been published. The renewed interest has been spawned by the entanglement of DNA strands and a desire to have a mathematical description of the knots which may form [11]. Aspect ratio is the ratio of the length of a body to its diameter. When this ratio approaches fifty or more, one of the requirements for "caking" begins to be met even if the body has no branching points. If the body has branching points, a new method of entanglement can result in a "brush heap" configuration that is more difficult to separate than "bird nesting," which will tacitly be assumed, are made from particles that are straight and smooth. Bird nesting could occur as a "log jam" when timber is floated downstream to a saw mill, or it could happen on a scale, as small as the formation of false gels when acicular crystals of colloidal dimensions precipitate to form systems with properties similar to gels [12].

In considering any class of caking, particle size and particle size distribution become important considerations. When everything else is equal, and it almost never is, the smaller the particle size of a sample, the more likely it is to cake. The reasons are obvious, but they should be reviewed. Firstly, the surface-to-volume ratio becomes greater as the particles become smaller. The contact area between particles increase and the surface that is free to adsorb water and gases increases. Conversely, the bulk density of very fine powders is usually greater than the bulk density of larger particles. This means that there is usually less free air or gas between the more tightly-packed particles. But, the bulk density also depends upon the true density of the particles, particle size distribution, their charge, aspect ratio, flexibility, branch points, and similar considerations. Bulk density is usually a rather arbitrary specification but can be of considerable value when confined to the context of the properties of a particular product. Bulk density will be considered in more detail in the section dealing with test.

Related to particle shape and mechanical caking is friction of the particles in passing over each other or the walls of a container. Friction is not an issue if the particles are branched or may become entangled. It is meaningful when considering particles where shape is not a mechanical barrier. Crystals that form plates such as kaolin, graphite, or molybdenum sulfide have low coefficients of friction, which is surely shape-dependent, but a plate-like shape does not necessarily mean that the friction if the particles is low. This is often seen with the abrasive in toothpaste and their dentin abrasion when the paste is brushed over a tooth. In this case particle size can be as important as shape is. If the particles are small enough, practically all substances show small abrasion indexes, but if the particles are more than twenty microns in diameter, many substances that might otherwise be expected to have low abrasive indexes can be very abrasive. The issue is involved and is not well understood, but friction can contribute to particle caking.

A very interesting positive application of mechanical caking has been published by Walmsley and Duffy. They employed mechanically entangled fibers as a medium for the hydraulic transportation of coal and iron sands in a pipeline. It is claimed that the fibers act as a drag-reducing agent and that the shear loss at the same loading is less than it is with water alone. The coal or iron sands are removed from the fibers by differential settling [13]. Although the Walmsley and Duffy did not make the claim, some of the author's twenty-five years experiences with tailings solids in mining operations suggest that this intriguing approach should also lessen the wear on pipelines. This can be a major expense in some hydraulic transportation lines.

Plastic-Flow Caking

Caking from plastic flow usually occurs with amorphous tars, gels or waxes. It is not required that the system be amorphous because some soft crystalline substances merge when subjected to either pressure or increased temperature. Even the sticking of adhesive tape could be classed as a type of plastic flow. As the name implies plastic-flow caking results because the particles have a yield value that is exceeded and the particles stick together or all flow into a single particle in the most severe cases.

Chemical Caking

Chemical caking is by far the largest subject to be considered. In this section chemical reactions such as decomposition, hydration, dehydration, atmospheric oxidation, and atmospheric carbon dioxide reactions are typical chemical reactions in which a new compound has been generated. Dissolution and recrystallization, solid-solid phase transitions and sublimation are typically physical in nature and may or may not involve the formation of a compound that is new to the system. What has sublimation to do with caking? It may be very important to systems that contain highly volatile solids. They may sublime and recrystallize between particles, gluing them together.

Usually physical change does not generate new compounds. When ammonium nitrate undergoes the phase transition at 32°C, it is ammonium nitrate before it goes through the phase transition and it is ammonium nitrate with a changed crystal form after it has passed the transition. This is not the case with the polyphosphate, misnamed insoluble sodium metaphosphate. When it is heated to the phase transition, around 450°C, it is an insoluble long-chain polyphosphate. But when it passes through the transition, it emerges as soluble sodium trimetaphosphate, a six-membered ring, with almost no similarity to the original compound. The mere fact that the insoluble sodium metaphosphate suffered a phase transition may well have produced a caked sodium trimetaphosphate. Whether or not it did form caked solids depends upon a number of factors too involved to discuss at this point. Sodium trimetaphosphate can pass through other transitions to new crystal forms without changing the molecular structure even though the crystal structure is changed.

Electrical Caking

Electrical caking is perhaps the most interesting form of caking and is a part of physics discovered by P. Curie. Volumes have been written on crystal structure and how this influences the properties of a crystal. The subject cannot be justly treated in this work. Hopefully the superficial treatment will lead the reader to explore this exciting area in greater detail.

Most crystalline substances are not symmetrical in their properties, either physically or chemically. The asymmetry allows the crystals to behave in interesting ways when they are perturbed by heat or stress or light or electrical fields or even sound. When crystals are symmetrical, they may be as beautiful as diamonds. They usually have little electrical functionality. They do not usually cake as a result electrical dipoles in the crystals. Static charges may cause troubles but problems caused by movements of charged bodies inside the bodies are unlikely. This is hedging because of electrets, which can be uncharged liquid waxes before they are solidified in a strong electric field, can exhibit charges, unrelated to piezoelectric behavior, although in some respects the behavior can be similar.

The four types of electrical behavior contributing to caking and flow problems are piezoelectric, pyroelectric, ferroelectric and static electrical behavior. The first three can be shown to be a result of crystal structure whereas static electrical behavior on powders is not nearly so easily predicted. Most powders are capable of becoming charged either positively or negatively. The study of static electricity is too often ignored as a "static field" and not enough attention is being devoted to the subject. In exploring static electricity in more detail and how it influences powders and surfaces, it will be shown that positive and negative charges do not behave in the same manner on surfaces and that a fundamental difference should be expected.

The electrical action expressed as a result of crystal structure usually exhibits its behavior under static conditions, while static charges are usually more pronounced under dynamic conditions. Pneumatically conveying powders can cause a variety of problems because of the static charges acquired by friction during the transportation process.

VARIABLES IN CAKE FORMATION

One of the prime difficulties in classifying caking problems stems from the very large number of variables that can contribute to caking. Fortunately, most of the variables can be identified readily and eliminated or controlled, at least in laboratory studies. In the section dealing with experimental design a few of the statistical methods of identifying and ranking variables will be discussed along with interactions. It should be expected that moisture and temperature can act together to produce more caking than either variable alone.

Composition

As noted, the interest is mostly in commercial products because it is doubtful that much interest can be generated in the caking of solids that are not of commercial significance. Formulated commercial products may range from pure substances, sugar for example, to very complex mixtures represented by food products or solid detergents. From a phase chemist's point of view ten or more components may be formulated into a single solid product. Minor changes in one or more of the components may cause dramatic changes in the product, depending upon the multi-coordinate phase diagram that is much too complex to construct.

Usually it is possible to isolate the major components in a complex mixture and to determine how the individual secondary components influence the major components. By a process of elimination a secondary component that is causing trouble can be isolated and perhaps a substitute can be found. An example of the types of secondary components that can cause troubles are those that have very high solubilities and are hygroscopic. Ammonium nitrate, calcium chloride, and tripotassium orthophosphate are the types of additives that might cause troubles in an otherwise free-flowing formulation. Once started it is almost impossible to stop the absorption of water from the atmosphere in any practical way.

While isolating variables it is usually helpful to divide the variables into three types: 1. Processing variables, 2. Raw material variables, and 3. Inherent chemical and physical variables. Examples of the variables are many and only a few will be mentioned. Processing variables may be improper temperature, feed rates, and weighings. Leaky trucks, rail cars, storage bins, or improper bags may also be included as processing variables.

Raw material variables depend upon the source of the raw materials and their handling while in transit. Many raw materials are mined products and are used without additional purification. Trona is a sodium carbonate mineral that is used in place of soda ash. It may contain all manner of trace elements.

Lime is another mined product that is fired and used without additional treatment. Of first importance is the impurities in the seam of limestone being mined. These impurities vary widely and it is good practice to explore many lime producers before choosing a lime supply. The firing of the lime is also important. There are two common processes. The best for chemical use is the vertical kiln. It is usually fired with either gas or oil. The gas-fired lime is usually the purest. Oil fired is next but the sulfate concentration in oil-fired lime is usually somewhat greater than in gas-fired lime.

The second type of firing of lime is rotary-kiln firing or cooking. The kilns used to fire lime and portland cement are similar. They are huge rotating tubes in which the temperature is maintained by adding the fuel at one end. In the lime and cement industry the rotary processes are usually co-current meaning that fuel and feed are fed into the same end of the tube. In the vertical kiln the reverse is true and the process is fed counter current with the feed being added at the top and the oil or gas fuel being added at the bottom of the kiln.

In the rotary kiln process, the cooking occurs by adding coal to the limestone and burning the two together. This means that the finished lime not only has all of the impurities that came with the limestone but also contains all of the ash and other impurities that were contained in the coal and did not burn to volatile byproducts such as carbon dioxide, sulfur dioxide, and water during the burning process. Considerable care should be exercised in using rotary lime in chemical processes. The rotary lime is satisfactory for most lime uses, but it is in the high-purity area that the impurities can cause problems. One well-known brand of toothpaste was experiencing troubles with the shelf life of the fluoride in the paste. The problem was traced to an impurity in the lime that was being used to make the paste. The impurity was reacting with and precipitating the fluoride.

It can be most difficult to deal with the inherent property of a substance. All too often the customer will desire that all other properties of the product remain unchanged but that the product no longer cakes. In purchasing a product the customer is usually purchasing a function or a property and does not really care what the particular chemical is as long as it safe, inexpensive, and does the job it was purchased to do. But, properties are inextricably tied to molecular structure and composition. If new properties are desired, a new molecule is usually required, although, sometimes new properties are obtained from admixtures. It should be recognized that in solving caking and other problems it is usually easier to find a substance to fulfill a use than it is to find a use for a new substance. A chemical with unusual properties may suggest its own use and it is for this reason that new compounds should continue to be prepared and tested for new uses.

SUBCLASSES

Certain properties of components as either pure substances or as part of a formulation can be expected to cause caking problems under certain conditions. A few of these will be mentioned.

Basic Components

Basic components are very likely to react with carbon dioxide of the atmosphere to form carbonates. Under these conditions a recrystallization process takes place, which can cause bridging between particles. Sodium hydroxide is a classic example. When solid sodium hydroxide reacts with atmospheric carbon dioxide, not only is sodium carbonate formed but water is also produced in sufficient quantities to cause the products to be damp.

$$2NaOH + CO_2 => Na_2CO_3 + H_2O \qquad [1]$$

Acidic Components

Acidic components are not as inclined to cake as a result of atmospheric gases, other than water, as are basic substances. When acidic formulations are used near livestock on farms, ammonia can become a concern. Large quantities of ammonia are generated from the waste of livestock. Reactions of ammonia with acidic products can cause bridging between particles, causing the product to cake. Acidic products are often formulated in conjunction with sodium bicarbonate in both pharmaceutical and food products. Baked products are likely to contain both acids and bicarbonates. It is necessary in both cases to protect the product from moisture.

Volatile Components

Volatile solids can sublime causing caking. These can range from ammonium halides, or ammonium carbonates to borate esters, or benzoic acid, camphor or menthol to mention but a few. In this case the solid can vaporize and recrystallize to form bridges between the particles of the formulation.

Low Melting Point Components

Low melting temperatures of solids can cause problems if the melting temperatures are 150°C or less. It is unlikely that most products should ever be subjected to temperatures as high as 150°C in storage but many substances can begin to soften many degrees below where they actually melt. Also in formulating it is possible to mix components that lower the melting temperature even more if the components already had a low melting temperatures when they were pure.

Unstable Components

Solids can be chemically and physically unstable for a wide variety of reasons. Several of the reasons have been mentioned as examples in other sections. The devitrification of a glass to a crystalline product or a transition of one crystalline form to a stable form may cause severe caking problems. Many chemicals exhibit disproportionation reactions. The reactions are more inclined to happen with oxidation and reduction reactions than with decompositions but monocalcium orthophosphate, Ca $(H_2PO_4)_2 \cdot H_2O$, is an example where the crystal decomposes to a mixture of dicalcium orthophosphate, $CaHPO_4$, and orthophosphoric acid. It is almost impossible to store the chemical without either caking or decomposition. As a result of this behavior, most substances labeled monocalcium orthophosphate have either an excess of orthophosphoric acid or an excess of dicalcium orthophosphate in them to stabilize the system. These chemicals should never be used in studies without first analyzing the samples, even when the label may state "Chemically Pure", or "Analytical Reagent Grade".

Oxidizing and reducing agents are inclined to decompose, particularly when in contact with each other. This is particularly true of bleaching agents, as peroxides, that are inclined to loose active oxygen and in so doing recrystallize to a caked form. Bleaching agents are often included in formulations and much care must be exercised in choosing compatible components. Not only are there the problems associated with caking, but fire or even explosions can result from improper choices when formulating. The decomposition of peroxides, in particular, is catalyzed by a number of inorganic salts as well as decomposing active organic compounds.

Weak Solid Structures

A problem can occur when the structure of a solid is too weak to withstand handling and packaging. Some substances grow as a dendritic crystal form. Sodium chloride grown in the presence of a small concentration of gum arabic exhibits a fern-like growth that is very fragile. It is possible to use crystal habit modifiers to change the morphology of crystals dramatically. Dyes are often used as modifiers. Some dyes will adsorb strongly to one face of a crystal and block growth on that face. This will cause the other faces of the crystal to grow more rapidly than the blocked face. The shape of the crystal can be grossly altered. Many times a much stronger crystal can be grown with habit modifiers in the crystallizers.

Crystal habit is not to be confused with crystal structure. Crystal habit is the shape of the external crystal while crystal structure is a result of the repeating configuration of the crystal lattice. The crystal habit of a crystal may be changed and the shape of the crystal will be unlike the crystals grown in other ways, but the crystal structure will remain unchanged. Another way to express this is to say that the x-ray pattern of the crystals will not be changed by a habit change nor will the angle between specific faces be altered by a habit modifier. The refractive indices of the crystal will also remain the same, although some of the faces may not be easily recognized. Phase transitions, on the other hand, change crystal structure but the habit may or may not be altered.

Usually there is a change of habit when there is a change of structure. An example of a change of structure without changing habit is exhibited by the dehydration of blocky dicalcium orthophosphate dihydrate to anhydrous dicalcium orthophosphate [14]. If carefully done the transition may take place with very little change in shape of the crystals. Additionally, the dicalcium orthophosphate may be heated to form calcium pyrophosphate, not only changing the crystal structure but also the molecular structure of the crystals while retaining the blocky morphology of the crystals. In each case the crystal structure has been changed and each crystal type exhibits a unique x-ray pattern.

Heat Capacity and Heat Conductivity

Heat capacity is important for a variety of practical reasons. The primary reason is that the higher the heat capacity of a solid, the more energy that must be supplied to the solid to raise its temperature one degree. If the solid could suffer a phase transition at a higher temperature, a high heat capacity may prohibit the temperature from rising to the required level at a constant heat load. Remembering that it is temperature that determines whether or not a phase transition occurs, a product stored in a cool warehouse may not have time to heat to a transition temperature before the daytime heat load is removed. Once the transition has been initiated, the energy input may be very large while the temperature remains constant as a result of the latent heat of transition.

Heat conduction can be just as important as heat capacity in protecting a solid from caking. If the conductivity is very poor, only the outer layer of a solid may be heated above the "critical" temperature at which the solid suffers a large change. It may also be that the transformed solid is a much poorer heat conductor than the product itself and that the outer transformation protects the bulk of the solid. Several studies have been reported in the older literature that deal with heat conduction and particle size. As expected smaller particles with more contact area are better conductors when they are well packed. As was also expected, the material the particles are made of also influences the heat conduction [15].

Most of the time it can be assumed that a temperature-induced change of a solid will begin at the outer layers of the bulk of solids and work its way into the interior, as the product heats. This does not necessarily have to be the case. Well-formed crystals, in particular, do not always react to a temperature if the system has not been seeded. The temperature may rise many degrees above a transition temperature and then suddenly an entire super-heated mass will react very quickly both on the surface and within the body of the contained solids. This can lead to caked materials and ruptured containers.

Seasonal Influences

Too often in the manufacturing process the influence of the seasons on a product is not recognized until a critical problem is confronted. Warm temperatures and high relative humidities can cause troubles that are not encountered in cold, dry weather. If problems are expected, it is possible to include preventive measures when the process is designed. Any time a product has high-solubility refrigeration equipment to lower the humidity of drying air may be considered. In one plant built in a swamp and intended to produce high-grade ammonium nitrate prills, the cooling air for the entire process was chilled to dehumidify the air but the process was economical to operate despite the cost of refrigeration.

The Kinetics of Cake Formation

Kinetic studies are usually made on systems with starting conditions that are relatively easy to reproduce. Gases, liquids, and solutions are the usual candidates, and it is, at times, very difficult to obtain reproducible data even from these systems, particularly if the data have been obtained at two or more different laboratories. Treating the caking of a solid under storage conditions by classical kinetics may be difficult to relate to the general science. One is probably just as well advised to treat the formation of cakes in a qualitative or semi-quantitative way even though the data may have been obtained under controlled conditions.

Cake formations of the type considered is not usually an instantaneous event. There are no reliable rules to predict how much time will be required for a product to begin to cake or how severe the problem will become. The only way to obtain reliable answers is by experimentation under conditions similar to those the product is expected to encounter. Many times worst-case conditions may be helpful in arriving at an answer.

It is probably a waste of time and resources to become too analytical in attempting to predict the kinetics of cake formation. The kinetics are of more laboratory interest than a marketplace consideration. Caking in the field is usually caused by some condition that is unexpectedly beyond control. It is unlikely that any product marketed today will cake under normal operating conditions. It is the goal of any study directed toward a caking problem to eliminate the problem entirely at least for the expected life of the product. Therefore, the value of any time study should yield information regarding the reliability of the laboratory test procedures and whether the laboratory test had been given sufficient time to give reliable answers.

Questions dealing with the aging, moisture absorption, and temperature can be addressed from a kinetic viewpoint but the value will be to establish maximum and minimum limits before a product is stored or shipped. Any time a powdered product is stored, either damp or hot, caking problems are more likely to occur. The time lost in handling or reworking an off-grade product is certain to require an excessive quantity of time and expense. It may destroy a customer's confidence in a supplier, placing an entire business in jeopardy!

COMPRESSION

There is a concept some plant personnel use in describing a caking phenomenon. It is called "compression caking" implying that the powder will cake if sufficient outside force is imposed on the powder. There surely is no doubt that most solids are more inclined to cake when subjected to compression from an external force. Not all substances are prone to caking because an external force is applied, provided the force is not of geological magnitudes.

If forces of the type encountered in a railroad car or silo are considered, they are seldom more than about 30 pounds per square inch, assuming a bulk density near sixty pounds per cubic foot. These would be maximum extremes. Because the force in a mass of crystals is supported by point contacts on the solids in a stack, it may well be at the point of contact pressures as great as 100 pounds per square inch may be encountered. The question of merit is, "Which types of solids cake under these conditions and which do not cake?"

The solids that do not cake under the pressures normally encountered in warehouses and railroad cars are hard crystals, strong crystals, crystals almost insoluble in water, chemically stable solids, smooth surfaced, blocky to spherical compared to acicular or fibrous solids, dense as compared to light, larger well-formed crystals compared to broken parts and smaller particles, symmetrical as compared to asymmetric, high-melting compared to low-melting temperatures. These properties are not usually influenced greatly by pressures in the range encountered in commerce.

When weights are added to samples in the more or less "standard" caking tests, the weights magnify the influence of gravity on the samples. It is usually expected that samples taken from the bottom of a stack or the bottom of a silo or railroad car are more likely to be caked than samples taken at other points. The reasons why pressure on a sample increase caking tendencies will be examined. This is in addition to simply causing the particles to be pushed closer together when they are under pressure. To explore the influence of pressure alone on a solid that does cake under pressure, sodium chloride is a good example to consider, because it has none of the physical and chemical properties that cause crystals to cake with the exception of solubility in water. It is assumed throughout this work that when the vapor tension of water in the atmosphere approaches the vapor pressure of a saturated solution of a salt, there is a saturated solution film of water and the salt on the surface of the salt.

It has been well established that systems that are inclined to cake under compaction and where solubility is a factor caking is much worse under hot, humid conditions. In most cases an increase in pressure on a saturated solution should increase the solubility of the salt. At the contact points between crystals, the pressure may become very high because the area of contact may be very concentrated. Any small cycling in temperature may then cause the salt to dissolve and recrystallize. After a few cycles the particles will begin to grow together while helping to relieve the stress imposed on the system by redistributing the load across the crystals. It may be noted that once caked, solids do not return to the original shape and volume when the load is removed.

Caking accompanied by loading with weights is seldom instantaneous but may require several days to approach some "equilibrium" condition. The end point may be the time required for all of the particles in a mass to cement into a single lump the shape of the container. The time required is an indication that the processes causing the caking are of the type that can be expected by a slow transfer of mass, probably confined to the surface of the crystals, unless the vapor loading is severe.

THE FLOW OF SOLIDS

In classifying the types of problems associated with powders, flow problems, even with systems that are not caked in the sense that formal bridges have formed between particles, have demanded much attention. This is probably a result of the large market for equipment designed to handle and store particulate solids ranging from powders to larger grains. Since the engineering aspects of bin design and powder transportation are important for the sale of heavy equipment, considerably more attention has been expended on mathematical treatments.

Three flow patterns are observed in the flow of solids from silos, bins, and hoppers. These are mass flow when all of the solids move simultaneously, funnel flow when there is a flow down the center axis of the container but the solids at the walls remain more or less stationary, and expanded flow [16]. When solids flow through a tube, air is also in motion in the tube. At the exit the pressure can be greater than atmospheric and the air in the tube may be flowing counter currently. This may cause arching or bridging of the solids in the tube [17]. Arches of considerable strengths may be formed in solid flow systems and may require external energy to break the arch and allow the solids to flow again. Systems flowing under these conditions may be erratic and may pulsate.

A.W. Jenike was responsible for bringing the science of solids flow to the attention it deserves [18]. He not only laid the basic science of flow but also contributed in a very positive way to the design of equipment to take advantage of the characteristics of the solids being handled. As is often the case in the design of equipment, it must be designed to do one particular job and it is not unlikely that the equipment will perform only marginally when applied to another solid with characteristics unlike the materials it was designed to handle. It can even become dangerous if the pressure on the walls becomes greater than the container can withstand.

It is not surprising that coal handling has attracted much attention. Huge quantities of coal are transported in automatic equipment. If the coal does not flow as desired, the equipment becomes inoperative and much of the advantage of automation may be forfeited in downtime. Other industries with similar problems and equipment are grains, cement, lime, ores, and clays.

In the chapter dealing with typical solutions, a case history is mentioned in which dramatic changes were accomplished in the flow properties of powdered limestone by a very simple and inexpensive treatment. This very simple solution was to grind the limestone in a very dilute solution of oleic acid. This illustrates that a change of equipment may not be the only option when dealing with a solid that is not behaving as desired in existing equipment. It may be possible to change the flow properties of the solid in a favorable, inexpensive manner. The first requirement is to understand what is required to make the solid behave as desired. There are a limited number of reasons that a dispersed powder flows sluggishly, if at all, under the force of gravity. In this case the discussion is not of powders that refuse to flow, but powders that refuse to flow in the time required to make a process profitable to operate. Much equipment is also sold to shake or vibrate containers to help powders to flow. These usually perform very satisfactorily but they are also very noisy. If other, even more satisfactory, approaches could be found for treatment of solids to improve flow and decrease the noise, operator fatigue would be decreased and a better product would be obtained.

The flow of powders is a digression from the main theme of caking and lumping but very little has been accomplished if a caking problem has been solved only to discover that the powder will still not flow from its container. Attention will continue to be directed more toward the laboratory than the plant and engineering aspects of the caking problems, but it is necessary that those involved with projects requiring a solution to caking recognize that equipment exists that can solve or improve the handling of solids. It is reluctantly that the flow of powders will not be given more attention. The published literature is rich and the subject is of great interest both theoretically and practically but most of the time must be devoted to caking and lumping, regardless of the flow conditions to which a powder may ultimately be subjected. In Chapter VIII, "Typical Solutions," the subject will be treated in more detail with literature references to many ingenious and successful applications of equipment design to solve caking and flow problems.

PRODUCT TYPES

One of the better reviews ever written on the caking of solids was limited to the caking of fertilizers only [19]. One of the classifications which must be noted from the article is product type. Not all products can be treated in the same way. It is obvious that this is the case but the similarities and differences can not be ignored. A box of detergent for home use and a warehouse of bagged fertilizers or of refrigerated cheese have many points in common yet they require their own special treatments as a result of the way they are used and marketed.

The fertilizer industry has developed many tests that are of value in predicting the behavior of fertilizers. Many of the tests rely upon compacting the powdered products for fixed intervals of time and then measuring the properties of the resulting cakes. Usually the strength of the cake is measured and is considered to be a significant result. These tests are particularly significant if a customer uses them as a basis for purchasing materials. More attention has been devoted to understanding why a cake will form under sustained pressure and less attention to the properties of an undesirable result. Both approaches have their strengths and weaknesses and both approaches should be used when it is obvious they are needed to eliminate a problem.

SUMMARY

There are four major classes of caking. They are mechanical, plastic flow, chemical and electrical. Examples of mechanical caking are wads of cotton, coat hangers, and brush heaps. Plastic flow occurs when substances with yield values low enough to flow contact each other and either stick together or merge to form one particle. Examples are tars and waxes. Chemical caking is the caking that involves all types of phase transitions that may occur from chemical reactions to hydrations or dehydrations. Examples range from detergents to fertilizers and food products. It is the most common of caking problems and usually involves water in some action. Electrical caking involves both static electric charges as well as the electric properties of unsymmetrical crystals. The piezoelectric, pyroelectric, and ferroelectric crystals are the type that can cake as a result of dipoles. Static charges can cause flow problems as well as weak cakes.

Subclasses of the four major types of caking are numerous and should be handled after the major classes have been identified. In all cases the identification of the class of caking does not solve a caking problem but does insure that one has approached the proper problem.

REFERENCES

10. Downton, G.E., Flores-Luna, J.L., and King, C.J., *Ind. Eng. Chem.* 21, 447 (1982).
    Karnaushenko, L.I., and Novichkova, T.P. *J. Eng. Phys. (USSR)* 51, 1090 (1986).

Tanaka, T., *Ind. Eng. Chem. 17,* 241 (1978).

Cotton, J.W., and Wheeler, T.A., *Chemical Reagents In Mining Processes,* Society Of Mining Engineers, Inc., Littleton, Colorado, p. 231 (1986).

11. Peterson, I., *Science News 133,* 328 (1988).
12. Levine, I.N., *Physical Chemistry,* p. 344, McGraw-Hill Book Company, Inc., New York, N.Y. (1978).
13. Walmsley, M.R., and Duffy, G.G., *Transactions of the Institute of Professional Engineers New Zealand, 14,* 101 (1987).
14. Griffith, E.J., and McDanniel, W.C., United States Patent No. 4,721,615 (January 26, 1988).
15. Partington, J.R., *An Advanced Treatise on Physical Chemistry,* Vol. III, p. 411, Longmans, Green and Company, London (1952).
16. Bussian, G., *Powder and Bulk Engineering, 2,* 36 (1988).
17. Heine, H.J., *Foundry Management and Technology, 116,* 34 (1988).
18. Jenike, A.W., *Bulletin 108,* University of Utah Engineering Experimental Station (October, 1960).
19. Bookey, J.B., and Raistrick, B., *Chemistry and Technology of Fertilizers,* edited by Sauchelli, V., Reinhold Publishing Corporation, New York p. 454 (1960).

CHAPTER
THREE

The Chemistry of Cake Formation

CHEMICAL AND PHYSICAL CHANGES

Most products that cause caking problems do so because of physical and chemical changes in the product. If a free-flowing powder were placed in a container and absolutely nothing occurred to the powder, it should remain a free-flowing powder until some change occurs. The mere movement of water from one phase to another when a product is stored in a warehouse or on a railroad siding in the hot summer sun may cause the product to turn to a brick or a dusty powder. All too often products are loaded into cars while they are many degrees above ambient temperature. In some chemicals this is likely to cause no problems while in others it can be a disaster for product quality. If the product is packaged above a transition temperature, the chances are very good that the product will emerge from the transition in a caked condition. This is a result of the fact that none of the crystals that were packaged above the transition temperature exist below the transition. Unless the crystals that were packaged hot have a delayed transition each of the original crystals will have crystallized to form new crystals. It is also likely that the original crystals transformed into many new crystals with a lower bulk density. This act will increase the force placed on the container, and consequently, the force exerted on the product itself.

The interaction of water with solids is the prime cause of caking in most instances. The problems caused by water may take a variety of paths, causing caking of several manifestations. Perhaps there is no better place to start than the solubility of a solid in water. If the solubility of a salt is known and its molecular weight is known, the expected behavior of the salt may be estimated in everything from a detergent box to a freight truck. Knowing the solubility, S_1, of a salt in water at temperature, T_1, and the solubility, S_2, at temperature, T_2, the solubility of the salt in water at any temperature between zero degrees and boiling is known by solving the simple equation:

$$\ln \frac{S_2}{S_1} = \frac{dH\,(T_2 - T_1)}{1.987\,(T_2 x T_1)} \qquad [2]$$

Where S is solubility in any convenient units but moles per 1000g. of solvent will cause other calculations to become more easily handled. dH is the differential heat of solution of the salt (at saturation). T is the absolute temperature. If one desires to use log rather than ln, divide the right hand side of the equation by 2.3026.

It is obvious that if one knows the solubility, S, at two temperatures, T, the dH term may be calculated and then the solubility at any temperature may be calculated. It must be recognized that the salt cannot change crystal form in the temperature range the calculations are desired. This is to say, if one hydrate exists at a lower temperature and a second hydrate exists at a higher temperature, one cannot expect the equation to accurately predict a solubility outside the temperature range of the hydrate stability. In any case this violates the definition of solubility. When two hydrates of a salt exist each hydrate has its own dH and it is for this reason that Equation 2 must not be used across a phase boundary.

The calculated solubilities obtained with Equation 2. are usually as accurate as the measured solubilities. The equation is remarkable in this respect. Not all results are so fortunate with most calculations dealing with colligative properties and caking.

The well-known expression for vapor pressure lowering in an ideal aqueous solutions is:

$$V_p = \frac{55.49}{nM_s + 55.49} \times V_{po} \qquad [3]$$

where the Ms is expressed in gram-ions per 1000g of solvent for ionized salts or moles per 1000g of solvent for neutral molecules.

V_p is vapor pressure and V_{po} is the vapor pressure of the pure solvent at the same temperature. Moles nM_s is equal to the gram-ions where n is equal to the number of ions into which the salt dissociates. In this equation it is assumed that all salts are completely ionized and have activity coefficients of unity. Since this seldom, if ever, is the case at saturation, the equation is at best an estimate but will yield at least an approximate answer if the saturated solution is not too concentrated. In the next paragraph the equation will be tested with an example.

Test the previous equation by calculating the relative humidity at which sodium chloride should begin to absorb water from the atmosphere at 25°C. Absorption of water from the atmosphere is the primary reason that sodium chloride cakes. The handbook solubility of NaCl is 36.1g per 100g of water. This is 361g per 1000g of water. Dividing 361 by 58.44 (the formula weight of NaCl) and multiplying by 2 (the number of ions in NaCl) one obtains 12.35 gram-ions at saturation, at 25°C. The vapor pressure of water at 25°C is 23.76 mm of Hg. Substituting the values into Equation 3, one obtains 19.43 mm of as the vapor pressure of a saturated solution of NaCl. If one now divides the 19.43 by 23.76 (the vapor pressure of water at 25°C) and multiplies this by 100 the relative humidity of the air in a closed vessel over saturated NaCl is obtained. This should indicate that any time the relative humidity of ambient air exceeds 82% at 25°C, NaCl shall absorb water from the atmosphere and it shall cake if the humidity falls below 82%. If the humidity remains above 82% the NaCl should deliquesce, leaving only a liquid.

All of this example is very simple theory but unfortunately it is not even close to being correct! As a matter of fact, a saturated solution of sodium chloride held at 25°C has a vapor pressure of 17.9 mm rather than 19.43 mm. The relative humidity over a saturated solution of NaCl is 75.3% and not 82%. Any time sodium chloride is exposed to air at 25°C having a relative humidity greater than 75.3% the NaCl will absorb water from the atmosphere. The reason for the very large divergence between the theory and the facts occurs because the activity of the sodium chloride in the solution has been neglected in the theoretical calculation [20]. This bad example is intended to clearly emphasize the need for measurements in dealing with caking problems. In the chapter dealing with test and test procedures, much attention will be directed toward obtaining data by measurement, particularly when they can be easily obtained directly from the system of interest. In most cases relatively simple tests and equipment will be very satisfactory. It is possible to obtain a relatively close estimate of the vapor pressure of a saturated solution of sodium chloride if all of the corrections are applied. But it is usually easier to obtain an exact measurement of the system than it is to go through the work necessary to calculate a close approximation.

Perhaps the footnote by S. Glasstone in *Textbook of Physical Chemistry* [21] on Page 651 best covers the state of simple calculations. "In reading this section (Osmosis and Osmotic Pressure) the student is advised to clear his mind of any theories concerning osmotic pressure which he may hold as a result of an elementary study of the subject." A similar statement is justified for students who have made an advanced study of this branch of physical chemistry. The usefulness of the chemistry is confined to dilute solutions and can be used as a guide in one's thoughts about the systems one may confront while dealing with caking problems but care must be exercised in utilizing specific calculations that have not been tested in the laboratory.

It is certainly obvious that the difference between 75% relative humidity and 82% relative humidity is a very significant difference in the properties of a product. In many parts of the world the relative humidity seldom exceeds 82% at 25°C. But, 75% relative humidity is encountered much more often. The manufacturer of a product that he felt was safe to 82% relative humidity is in for a shock when he discovers that his product absorbs water at a much lower water vapor loading. It should also be noted that almost any soluble impurity in the product should increase the chances of water absorption at relative humidities even lower than 75%. It will be seen from typical phase diagrams of these systems, that most mixtures have a combined solubility greater than either pure member and this should cause the system to absorb water at vapor loadings less than either pure substance alone, provided there is neither solid solution nor compound formation [22].

Figures 1, 2 and 3 demonstrate what happens if a salt is first exposed to a vapor tension above the vapor pressure of a saturated solution of the salt. In this case the salt is prilled ammonium nitrate.

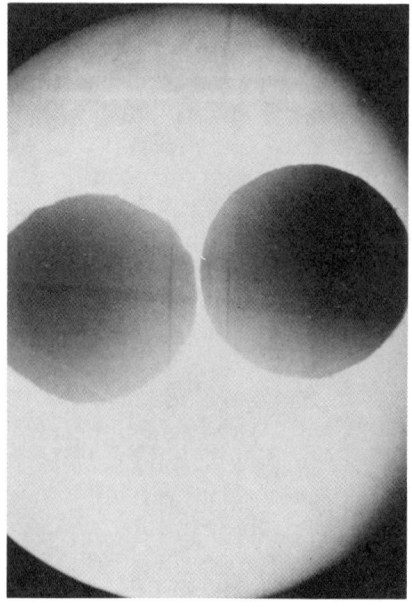

FIGURE 1. *Microscopic view of two prills of ammonium nitrate (50x).*

Chemical and Physical Changes / 37

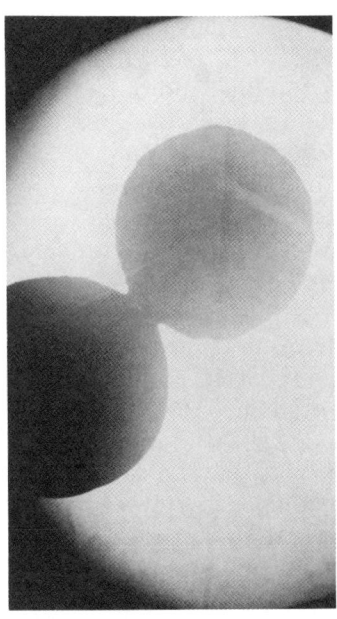

FIGURE 2. *The prills of Figure 1 after being exposed to a high humidity of human breath for two minutes.*

When exposed to the high humidity, the salt begins to deliquesce on the surface. Next the salt is transferred to an atmosphere where the vapor tension in the air is much less than the vapor pressure of a saturated solution of the salt. The water in the solution on the surface of the prills evaporates. This allowed the salt to crystallize from the

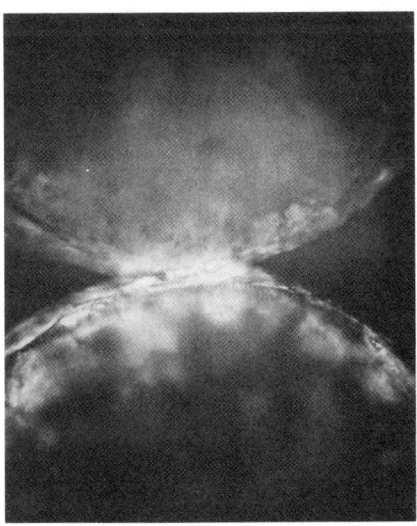

FIGURE 3. *The bridge that formed between the prills of Figure 2 when the moisture evaporated. Polarized light (100x).*

saturated solution on the surface of the crystals. As the salt crystallizes the crystals bridge between the particles, cementing them together. Obviously any product subjected to this kind of treatment should form a very strong cake. Ammonium nitrate as sodium chloride has no hydrates at room temperature and the liquid quickly collects on the surface of the prill. Whether or not a salt forms hydrates will make a very great change in how it reacts to water vapor.

In Chapter IV it will be seen that Figures 7, 8 and 9 are similar to Figures 1, 2 and 3 except that the salt is heated through a phase transition. In this case all that happened was the crystals were forced through a solid-solid phase transition. The result of the bridging was very much the same in both cases but the mechanism for forming the bridge is completely different. As will be noted, the ions in ammonium nitrate sublime or distill from one crystal to another. It is difficult to determine whether there is or is not any instance when there is a liquid phase. The behavior of additives to the salt suggest that it is probably sublimation from one phase to the other.

Water vapor has some interesting and often unpredictable influences on crystals. The temperature of solid-solid phase transitions can often be changed many degrees depending upon the vapor pressure of water in the systems to which they are exposed. (See Figure 14 in Chapter IV.) In other cases hydrates are easily dehydrated in humid atmospheres, while in nearly anhydrous conditions much higher temperatures will be required to dehydrate a salt [23]. It has also been reported that when sodium tripolyphosphate hexahydrate is dehydrated under normal atmospheric condition the anion is destroyed but if the salt is dehydrated under conditions where the vapor pressure of water is in excess of 500 mm, the salt will give up its water of hydration without harming the anion [24]. Two factors are probably active in the behavior of the water. It is surely changing the surface energy of the crystals and it is acting to plasticize the solids. It will be shown in the next section on hydrates that phase transitions, such as dehydrations, usually if not always, begin at some point of strain on the surface of a crystal. The point of strain may be an intersection of faces or a flaw in a face. The more imperfect the crystal the easier it is to dehydrate in most cases.

In most of the work to follow, simple questions that may be answered "yes" or "no" will be asked, when possible. It is usually not too difficult to obtain "yes" or "no" answers to questions. The difficulties may arise in obtaining quantitative answers. But, as is demonstrated by all electronic calculators and computers, enough "yes" or "no" answers become quantitative when properly sequenced. Computers and modern calculators respond to binary arithmetic and symbols.

Some very complex mathematical approaches to the caking strengths of NaCl have been published. Tanaka published an article in which he employed 42 alphabetical characters to express the caking of solids by bridging [25]. His data included work with NaCl, KCl, $(NH_4)_2SO_4$, $NaNO_3$, and urea. His data fitted the predicted curves well. He concluded that temperature, humidity, mass transfer, and structure were important to the caking strengths of the solids. The work has some interesting ideas concerning evaporation and dissolution but cannot be generally applied to systems that form hydrates or suffer phase changes, without introducing even more variables. For those interested in caking modeling, the paper is worthy of study and is recommended.

WATER

It is absolutely necessary to comprehend many of the properties of water if the caking of commercial products is to be understood. Commercial products are singled out because there is seldom a practical way to protect large volume products from the water in the atmosphere during manufacture and delivery. The following table lists some of the important properties of water as they relate to caking. There are many other properties that might be mentioned that range from the structures of ice and water to the phase diagrams that have been obtained at very high pressures and extreme temperatures. Although these properties are very interesting, it is difficult to relate them to mundane caking problems.

The values in Table II should be studied until the range of the variables is familiar to anyone working with the caking of water soluble and hydratable solids. This does not mean that they should be memorized but merely some awareness of what to expect from changes that may occur in a system because of the interaction of the system with water. The values for the vapor density as a function of temperature were presented primarily to show how much change occurred with saturated steam as a function of temperature. These values are not the air saturation values but the quantity of water that would fill an evacuated cubic meter at the specified temperature and are, therefore, the pressure of water and not the partial pressure. As adsorption of water onto the surfaces of crystals and amorphous solids is considered, an understanding of these values and how they relate to temperature will be helpful.

Water is involved in some manner in more than ninety percent of all caking problems and should be the very first substance to consider when dealing with a caking problem of unknown origin. There

TABLE 2   The Properties of Water that Relate to the Caking of Solids

Property		Handbook Value
Molecular Weight		18.02 g/mole
Freezing Temperature		0.000°C
Boiling Temperature		100.00°C
Density of Water 0°C		0.99987 g/ml
Density of Ice 0°C		0.917 g/ml
Heat of Fusion		79.71 cal/g
Heat of Vaporization		539.55 cal/g
Specific Heat 25°C		0.99828 cal/g
Viscosity 0°C centipoise		1.7921
Surface Tension 25°C		71.97 dyne/cm
Index of Refraction		1.33241
Dielectric Constant 18°C		81.07
Specific Resistance		($2.5 \cdot 10^7$ ohms/cm)
Cryoscopic Constant		1.86°/mole/1000g H_2O
Ebullioscopic Constant		0.52°/mole/1000g H_2O
Coefficient of Expansion	20°C	$0.207 \cdot 10^{-3}$cc/cc/°C
Vapor Density	10°C	9.40 g/m^3
	20°C	17.28 g/m^3
	30°C	30.36 g/m^3
	50°C	82.85 g/m^3
	70°C	197.9 g/m^3
	90°C	423.2 g/m^3
	100°C	597.4 g/m^3

are many properties of water that relate to the caking of solids that are more or less obvious. Many of these are discussed in other sections. But there are many properties that are not obvious. Water has a remarkable influence on the Form IV—Form III transition of ammonium nitrate, little more than trace quantities will lower the temperature of the transition many degrees. The dehydration of dicalcium orthophosphate dihydrate, $CaHPO_4 \cdot 2H_2O$, (Yes! the dicalcium phosphate contains only one calcium ion per phosphate anion.) to anhydrous dicalcium orthophosphate, $CaHPO_4$, is catalyzed by water vapor [26]. Also it is known that sodium tripolyphosphate hexahydrate, $Na_5P_3O_{10} \cdot 6H_2O$, can be dehydrated to Form II sodium tripolyphosphate, $Na_5P_3O_{10}$, without destroying the anion provided it is dehydrated in an excess of water vapor [27]. Under most conditions when sodium tripolyphosphate hexahydrate is dehydrated, the anion is decomposed to a mixture of pyrophosphates and orthophosphates [28].

Some of the properties of water that cause it to have an influence on caking are the solvent properties, the solvation properties, the dielectric constant, the extremely high heat of vaporization, the high heat capacity in both the vapor and liquid states, its response to colligative property changes, particularly osmotic pressure and vapor pressure changes that aid water to penetrate many barriers, the surface tension, the liquid and vapor densities, the viscosity, conductivity, both heat and electrical, the absorption spectra from microwaves to infrared and visible, and even the refractive index to some degree. Each of the properties that have not been previously considered will be discussed and how they influence caking will also be noted.

The high dielectric constant of water causes it to have a pronounced influence on systems that contain ions. The dielectric constant is a contributing factor in its property as a solvent for salts. Because the solutions of salts are conductive of electricity, it allows large charges to accumulate on surfaces if the surfaces cannot be grounded. If the systems can be grounded, they form a path for the discharge of the static charges but not the ionic charges, which will behave as double layers at the surfaces.

The heat of vaporization is related to the free energy of vaporization, which in turn contributes to the free energy of the following reaction when A is hydrated with vaporous water.

$$mA + nH_2O \Rightarrow A_m \cdot nH_2O \qquad [4]$$

whereA is any hydratable molecule or salt.

Both the free energy of hydration and the free energy of vaporization (condensation) drive the reaction and as the phase change takes place the system can form cakes.

The high heat capacity of water is carried over into the hydrates and may be approximated by Kopp's Law. This means that the heat capacity of a hydrate is usually greater than the salt from which it formed and because the heat capacity of water is high, the more water molecules contained in a hydrate the greater the heat capacity of the hydrate becomes. The specific heat will usually become greater as well. The density of hydrates is usually less than the anhydrous salts from which they are formed. This means that a mass of salt will resist temperature changes as a hydrate more readily than the same mass of salt in the anhydrous condition. This can help to protect a salt from caking under conditions where the hydrate water is held very tightly, but not for loosely held hydrate water. The hydrate can be a better conductor of heat than the anhydrous salt and in some cases the conduction may overcome some of the advantage of having a higher specific heat.

Both the lowering of vapor pressure and osmotic pressure changes may cause problems with packaged foods that are hygroscopic or highly soluble. Samples of ammonium nitrate prills have been observed to absorb vaporous water— to deliquescence in small heat-sealed envelopes of three mil polyethylene. Vapor pressure lowering and how it influences caking has been discussed in several sections of this book.

Surface tension and viscosity play a role in the surface wetting of particles. The surface tension of water can be great enough to inhibit water from penetrating crevices between crystals in a particle. Most dissolved inorganic salt act as body active agents as contrasted to surface active agents. In other words the salts usually cause the surface tension of water to rise rather than to decrease. The higher the viscosity and the surface tension of the solution, the more resistant it will be in penetrating crevices. If the salt is very water soluble, the liquid will probably require much more time to wet surfaces and penetrate crevices.

Heat conductivity is more important to crystals undergoing phase transitions while electrical conductivity is more important where electrical charges are involved. In many cases electrical conductors are also good conductors of heat. A poor conductor of heat may protect a product from temperature fluctuations, the outside of the mass acting as insulation for the inside of the mass. If a piezoelectric crystal can be made to be conductive, it is protected from caking by dipole interactions.

If a salt absorbs radiant energy, the energy will cause the salt to heat. This can happen under conditions where the ambient temperature is low if the product is held in an insulated area. Conditions of this type are not often encountered with most powders of commerce.

ADSORPTION

Adsorption is a surface phenomenon, while absorption is a bulk phenomenon. The moment a gas or liquid begins to penetrate into a solid it is absorption. It cannot be assured that the following interpretation of the action of water with a surface is entirely correct; appropriate skepticism is advised. It has yielded results that have been successful in solving a variety of practical problems even if some of the details may be incorrect.

Adsorption is usually divided into two types. One type is called physical adsorption while the other is called chemisorption. Adsorption in the sense employed here is referred to as the sticking of a gas to a surface, either solid or liquid, when a clean surface is presented to the gas. In particular it will refer to water vapor and its interaction with surfaces.

Physical adsorption is classed as the adsorption of a gas on a surface where the gas is held by Van der Waals' forces. This means that there is little or no chemical reaction between the solid surface and the higher the pressure of the gas the more that will stick to the surface at a fixed temperature. When the temperature is held constant as the pressure is increased, the results are referred to as adsorption isotherms. With respect to the influence of water vapor on the caking of a solid, theory would indicate that the greater the partial pressure of water vapor in the atmosphere, the greater the quantity of water adsorbed to a surface. If this is put to the test, it will be found that gas will pile up on a surface in at least five different ways depending on the nature of the surface, but in general, the greater the pressure of water vapor in the atmosphere, the greater the quantity of water adsorbed. The difference between physical adsorption and chemisorption is that physical adsorption is reversible. If the partial pressure of the gas in question is decreased, the quantity of gas held by the surface is decreased and this action is reversible.

Physical adsorption is relatively well understood, and it is probable that just about everything in the atmosphere of Earth at normal ambient temperatures has at least a monolayer of physically adsorbed water on its surface. The subject of chemisorption gets a bit more difficult to define. First of all, the desorption is frequently not reversible and when the partial pressure of the adsorbed gas is decreased over the surface, the quantity of gas adsorbed may not change unless the temperature of the system is increased or a second gas is introduced into the atmosphere that competes for the surface more successfully than the adsorbed gas.

A second characteristic of chemisorbed gas is that it may have an activation energy to function. This means that the temperature might have to be raised to start the adsorption process but once started the heat of adsorption may be great enough to cause the action to continue. It is usually considered that the heat of adsorption of a gas by chemisorption is much greater than the heat of adsorption by physical adsorption. The heat of physical adsorption may be but slightly greater that the heat of vaporization (heat of condensation) of the gas at the specified temperature. This should give us a reasonably good idea of what to expect without spending much time with the named adsorption isotherms as, Langmuir, Freundlich, and so forth, which are presumed to have some serious flaws anyway. As in most cases the treatments have been an attempt to explain complex physical behavior with simple mathematical models. This approach will be avoided as applied to caking problems. The intentions are to build on the concepts brought forward in these works and, although they are not perfect that should not distract from the great value of the works.

In the following discussion the remarks will be confined to water vapor, but it should be understood that carbon dioxide, hydrogen sulfide, or just about any other gas could be considered provided the gas has similar interactions with the surfaces under consideration. In Table II it was shown that water has a very high dielectric constant. This means that it is highly polar and this will influence its behavior on a surface in many ways including the solubilization and ionization of salts.

The following types of surfaces will be considered:

1. Sparingly soluble crystalline salts.
2. Soluble crystalline salts.
3. Soluble amorphous salts.
4. Sparingly soluble amorphous salts.
5. Sparingly soluble amorphous neutral solids.
6. Soluble amorphous neutral solids.
7. Sparingly soluble crystalline neutral solids.
8. Soluble crystalline neutral solids.

There has been a distinction made between salts and neutral solids. For example sodium chloride is considered to be a soluble crystalline salt, while window glass is considered to be a sparingly soluble amorphous salt. Sugar is a soluble crystalline neutral solid but is not a salt. Each of the surfaces react differently with water and their inclination to cake also varies. It is not differentiated as to whether the surfaces were capable of forming hydrates with the vaporous water. This will introduce one more complexity into the study but it will probably depend upon whether the water is physically adsorbed or chemisorbed. Chemisorption will probably result in hydrate formation in most cases, but in the case of soluble amorphous salts, such as sodium polyphosphate glasses, it may be very difficult to distinguish adsorption from absorption. Yet another issue that will enter is low energy surfaces, as found in many polymer solids, and how water vapor will interact. Problems are found both in handling and transporting these low surface energy solids, and the influence of water is not clearly defined (e.g., nylon can contain more than 10% water).

Each of the eight cases listed previously will be discussed. It will be assumed that a "clean surface" is presented to an atmosphere of water vapor in air at 25°C and 80% relative humidity. The reader should determine what the vapor tension of water is under these conditions. The concept of a clean surface can cause much consternation, or even a surface that is not well defined for that matter. Assume that a clean surface is one on which no water molecules can be detected by whatever test that could be chosen. The surface is free of water. If a gaseous molecule of water collides with the surface and does not rebound, it has at a minimum given up the energy of vaporization and this will cause the temperature of the solid, and ultimately the vapor, to rise by the average translational energy of a water molecule at 25°C. Not all water molecules have the same energy in the vapor. It was Maxwell who clearly brought this point forward [29]. The surfaces undoubtedly capture the lower energy molecules at least at the beginning of the process. On surfaces that form hydrates it is necessary that the low energy molecules commence the process of hydration or the energy will be too great for the molecules to remain fixed to the surface.

Case 1.

Water is absorbed on a sparingly soluble (insoluble) salt. This may be, for example, anhydrous calcium sulfate. Calcium sulfate forms several hydrates and may become severely caked if left open in a humid environment as anyone who has done much work with plaster of Paris ($CaSO_4 \cdot 1/2H_2O$) may attest. If the water not only adsorbs to the surface but forms either the half hydrate of plaster of Paris or the dihydrate of gypsum not only the heat of vaporization is liberated but also the heat of hydration. If the salt had been soluble, a new complication should have arisen because the heat of solution would have been endothermic and this would have caused an opposite heat action to the ones just described.

Case 2.

The soluble crystalline salt surface is one with a number of potentially different actions as it reacts with water. It is important whether the salt forms a hydrate and additionally, how soluble the salt may be at 25°C when the 80% relative humidity that is imposed, is considered.

It is important to consider both whether the salt is capable of forming more than one hydrate and whether the salt is already hydrated, the water being a part of the crystal lattice when introduced to the vapor. One physical chemistry text book considered the vapor phase hydration of anhydrous $CaCl_2$ as absorption. (Lavine pg. 336) If the process were allowed to precede to complete hydration, this should be true but if the vapor tension over the $CaCl_2$ is so low that only surface water is absorbed then, it may be considered either chemisorption or even physical adsorption if the water can be removed reversibly. Calcium chloride is an extreme example because it is an excellent desiccant. Magnesium sulfate might be a better choice because it is not nearly so soluble as calcium chloride but does form hydrates.

Soluble crystalline salts that do not form hydrates or hydrated salts in their highest hydrated crystalline form at the chosen temperature behave similarly. These salts can form saturated solutions which may or may not be concentrated enough to continue to absorb water from the atmosphere. They will exhibit first the liberation of the exothermic heat of vaporization followed by the endothermic heat of solution of the salt. This variety of salt is the type most likely to cake as a result of high vapor tensions of water in the atmosphere.

Case 3.
Soluble amorphous salts present a difficult surface to classify. The amorphous salts are sometimes referred to as infinitely soluble, because a single phase system exists all the way from an anhydrous solid to an infinite dilute solution. In other words water and the solid salt mix in all proportions and that there is no discontinuity in the system from one extreme to the other. An example of this behavior is a soluble polyphosphate glass to which water is added. The glass is a supercooled liquid, and when a small quantity of water is added to the glass, the water and the glass mix. More and more water may be added until an infinitely dilute solution is obtained.

When a water molecule contacts the surface of a soluble amorphous substance, the molecule of water is adsorbed or absorbed depending upon how the system is viewed. The surface is not a thermodynamically defined system. Equilibrium is never established, but it is possible to equilibrate a system to the degree that an average condition may be established. Even the molecules in a system of this type are an array of molecular types; soluble phosphate glasses, for example. It is also possible to obtain glasses as the case of magnesium nitrate-ammonium nitrate glasses discussed in Chapter IV where the molecules are all of the same or similar kind.

Unlike the sparingly soluble amorphous salts mentioned in the following case, water has a drastic influence on the caking characteristics of the water soluble salts. The surfaces can become spongy and sticky in some cases, while the salts can crystallize in others. Laboratory test should definitely be made on these systems to determine how they will react to any set of conditions to which the salts may be subjected. It has been learned that additives that lower the rates of solution of these salts usually improve their performance in caking tests where hygroscopicity is a factor.

Case 4.
Sparingly soluble amorphous solids are probably influenced little more than having a layer of randomly adsorbed water molecules on

their surfaces. Glazed ceramics and silicate glasses are the types of surfaces under consideration. It is not at all clear what a surface is when sparingly soluble amorphous solids are explored. More and more is being learned about the crystalline surface with the advent of the tunneling microscope and other new tools. It is now seen in detail pictures that the surface is unlike the conventional unit cell of x-ray crystallography. The surface unit cell is rearranged and is larger than the unit cells of the lattice but the cells are still a part of an orderly array and the structure is very evident [30].

Orderly structure cannot be a characteristic of an amorphous substance. The water molecules are probably held at random sites, which have no reproducible property except in a statistical way. Fortunately, for this work these substances are not inclined to cake badly because their solubility is so low as to be considered insoluble and their properties are usually symmetrical to the degree that electrical behavior other than static charges can be ignored. They can be considered as chemically and physically sterile under most ambient conditions from this perspective. These solids may be considered as suspended in time, awaiting some long-delayed crystallization event, that must itself await a very slow molecular disentanglement process. The water is probably physically "sorbed" on these surfaces and is semi-reversibly desorbed. The desorption is very slow to reach an equilibration but not a true equilibrium state because the glass surface is not a thermodynamically defined state and cannot engage in an equilibrium in the same sense that a crystalline solid can participate.

Case 5.
Sparingly soluble amorphous neutral solids may include substances— such as, tars, gums, rubbers, many polymers and plastics. The composition of the systems will do much to influence the interactions of water and the solids. Neutral solids rich in fluorine are very difficult to wet and water is expected to be held very loosely to these surfaces. In this instance water can even help to prevent caking.

Case 6.
Soluble amorphous neutral solids present many of the properties of the soluble amorphous salts. These include some of the water soluble polymers as polyglycoles, polyvinyl alcohol, methylcellulose,. They adsorb, absorb, and imbibe water. Water acts as a plasticizer on most of these systems and sufficient quantities of water may cause them to have a caking problem but under most conditions water vapor does not seem to create many caking problems unless the conditions are extreme. The water is most likely physically adsorbed and absorbed and will probably release a part of the water reversibly. Long periods of time may be required for the systems to equilibrate and equilibrium

in the thermodynamic sense is unlikely to occur. These systems may be referred to as exhibiting floating thermodynamics.

Case 7.
Sparingly soluble crystalline neutral solids may include substances oxamide, solid paraffins, camphor, menthol, thymol, silica, elemental phosphorus, sulfur, gold, clays, and many more elements and compounds. The organic compounds and the nonmetal elements may have low-energy surfaces. Although they all probably adsorb some water on their surfaces, it plays a very small part in their tendencies to cake in many of the substances. There are many exceptions, however, and clay is a good example. Plastic flow, transitions, shape, and electrical charges are much more inclined to be involved. The water that is adsorbed is probably reversibly desorbed in most cases unless the compounds contain highly polar groups.

Case 8.
Soluble crystalline neutral solids are plentiful and include substances as sugar, urea, paraformaldehyde, acetamide, urethan, resorcinol, and thousands more. These systems are similar to Case 2 in most ways except that the salts are ionized while the neutral crystalline solids usually dissolve without molecular changes occurring. The vapor pressure lowering is usually less for neutral solids than for most ionized salts. This will usually lessen the tendency for the solids to deliquesce and changes in atmospheric conditions may not be as severe.

Case 9. (The solution surface)
When a solution is concentrated enough for the water of the atmosphere to be absorbed and to dilute the solution, the transfer of water molecules to the surface is merely greater than the departure of water molecules from the surface. The energy effect by the capture of a molecule is the same as it is on a solid except the two thermal activities are the conversion of the kinetic energy of the water molecule to heat and the heat of dilution of the solution. The heat of dilution will be meaningful only when all of the solid phase has disappeared. Otherwise the increase of water to the solution would cause more of the solid phase to dissolve. In most caking problems this is the case and more and more solution builds up on the surface of the solids.

A condition that is too often ignored when considering caking is the adsorption of ions on the surface of a crystal and the charge this imparts to the surface of a crystal. Colloid scientists and analytical chemists have known for many years that when a substance is crystallized from a solution, the crystals will preferentially adsorb its own ions from solution. This concept is very important in the manufacture of photographic films. Assume that one desires to precipitate silver halide from solution. The properties will vary grossly depending upon whether silver nitrate is added to a solution of sodium halide or sodium halide is added to a solution of silver nitrate. If the sodium halide is added to the silver nitrate solution, the precipitate will adsorb silver ions on to its surface and the crystals will have a positive charge. If, on the other hand, the precipitation is reversed and the silver nitrate is added to the sodium halide solution, the crystals will adsorb halide ions and the crystals will be negatively charged.

It should be expected that such a small difference in which solution was added to the other should make small difference. It turns out that the positive crystals grown from an excess of silver nitrate are much more active photographically than are the negative crystals. It is rather easy to understand why this is true when it is considered that a photon of light first reduces a silver ion to form a latent image that is ultimately reduced with the developer. Halide ions are not active in this manner but it is known that iodide is more active than bromide, which is more active than chloride.

In the early days of photographic films, they were very low speed and egg white was often used to form the emulsions in which the silver halide was suspended before it was put on a glass backing. In time it was recognized that the sulfide in egg white was reacting with some of the adsorbed silver ion on the surface of the crystals of the silver halides, converting them to silver sulfide. When gelatin became the carrier of choice, the speed of the film increased even more.

It is easily seen why the way a crystal is grown may have a very large influence on whether or not it is inclined to cake. If a part of a charge in a crystallizer is filtered while the solution is rich in cations and the second part of the crystallization takes place in a excess of anion precipitating agent, the two crystals will be of opposite electrical charge and will be inclined to cake. Order of addition not only influences the charge of a crystal but may make gross changes in the morphology of the crystals. Calcium orthophosphates are extremely susceptible to the order of addition of reagents [14].

COLLOID SCIENCE

Colloids probably play a much greater role in the properties of solids and how these solids behave than they are usually given credit for contributing. No discussion of surfaces would be complete without a discussion of colloids and their potential use. Colloid science has been used in an oblique way throughout these discussions but the importance of the surface chemistry and physics cannot be over stressed. At first glance the mere fact that a colloid is nothing more or less than a particle size range of any kind of matter ($1 \cdot 10^{-4}$ cm to $1 \cdot 10^{-7}$ cm in diameter) should make them so special. A colloid is a particle that is too small to be a suspension but too large to be an ion or most molecules. One of the properties that makes colloids so special is their electrical charge. The interest is as to how these charges interact with surfaces to change the properties of the surfaces as they interact to form cakes.

Monofilms have been discussed often as though the subject were new. Benjamin Franklin demonstrated in London that a spoon full of oil would spread over more than an acre of lake and that the thickness of the oil could be no more than a few angstroms. Today it is known that the layer was about one molecule thick. When substances, such as oleic acid, can improve the flow properties of insoluble solids, such as limestone, by grinding the limestone in a dilute solution of oleic acid, it was recognized that the influence can be gained with only a monofilm on the surface of the system.

When fertilizers as urea and ammonium nitrate are treated with clays, the clays contain dust of the colloid dimensions. It is probably true that the dust has as much to do with the reduction of caking as the bulk of the clay. This suggests that the smaller the particle size of the clay the more efficient it is likely to be. One must take care not to create a dust problem in the quest for higher efficiencies of clays. Many clay products are deflocculated with polyphosphates, soluble silicates, or sodium hydroxide in order to yield small free particles of the clay, and it is the particles in the colloidal size range that deflocculate. Kaolin slips (dispersions) have been maintained in suspension for several years when deflocculated with sodium tripolyphosphate [31]. Colloidal dispersions of calcium alkyl benzene sulfonate are also stable for years when properly prepared. These colloids should be superior to surfactants on coatings as those used on ammonium sulfate to keep it from caking. It has not been possible to determine how much work has been done with hydrophobic colloids specifically designed to eliminate caking problems. Emphasis has been placed upon work with hydrophilic colloids, particularly in the food and pharmaceutical industry, to eliminate caking and sticking problems. The substances were or could be peptized to colloids but this was seldom mentioned. Much of the chemistry of gums, starches, sugars, proteins, and water soluble polymers may be classed as colloid science, and their properties in both preventing caking or causing agglomeration is directly related to their colloid function.

HYDRATES

The formation of hydrates remains unpredictable and no satisfactory theory of why salts form hydrates or which hydrates form, if they do, has come forth. Whether an untested salt will form a hydrate, how many hydrates will form, and how many water molecules will be involved in each hydrate or the range of stability of the hydrates must all be determined analytically. These factors can become very important in the caking of crystals. On the positive side, the formation of hydrates gives water-holding capacity to many types of products. The product may contain or be subjected to water vapor without becoming wet. If water is held loosely in the crystals, some crystals will dissolve in their own water of hydration as the temperature is raised. Other crystals lose their waters of hydration as the temperature is raised and will then absorb it again as the temperature is lowered. If there is any way that a product may cake, it is more likely to do so when subjected to temperature cycling. The statement can be made even stronger. If there is any way a product can cake, it will cake. Most of the time products inclined to cake must be prevented from doing so.

All hydrates exhibit a vapor pressure of water, which changes as a function of temperature. In dealing with hydrates and mixtures of hydrates and anhydrous salts, it should be remembered that the change of vapor pressure is not a continuous function of temperature alone. Composition is also a variable. Depending on the particular phase diagram involved, the vapor pressure will often remain constant at a fixed temperature, even though water is lost from the system, as long as the equilibrium phases continue to coexist. A change in vapor pressure may occur abruptly when a phase is either gained or lost.

Consider a salt that has but one hydrate and when heated dehydrates to anhydrous crystals. If the hydrate is heated to the temperature at which it begins to lose water, an anhydrous phase is formed in situ with the hydrate. As more energy is added the system will attempt to resist change. The loss of water from the hydrate will require energy and this vaporization of water will keep the temperature constant, refrigerate the system, and, therefore, the vapor pressure will continue to remain constant, provided energy is not added to the system more rapidly than the system can respond. As more heat is added to the system, the loss of water from the hydrate will continue until all of the water is lost and the vapor pressure abruptly drops to zero. A similar event will occur if the salt exhibits two or more hydrates. The difference being that when all of the higher hydrate is lost, the vapor pressure will decrease to the vapor pressure of the less hydrated form at the temperature at which the last of the more highly hydrated form disappeared. It is perhaps obvious, but should be mentioned, that when a salt forms multiple hydrates, the most highly hydrated salt is always the lowest temperature salt, and the lowest hydrate or anhydrous salt is always stable at the highest temperature.

It is recognized that in the above discussion a multi-dimensional space is considered. The variables are composition, at least two phases, temperature, and pressure. In the ideal case one considers equilibrium conditions. For any experimental observation this is impossible if conditions are changing. An added complication occurs as a result of the rate of change and how far from equilibrium the experiment has strayed. If either temperature or pressure is held constant, then conditions can approach the ideal if the changes occur very, very slowly. Even so, it is seldom that the abrupt changes shown in most textbooks ever occur. Usually the changes from one condition to another happen as a smooth continuous function or an erratic unstable change, rather than a sudden, right-angle discontinuity.

As the temperature is raised, some crystals hold the water tightly, while others lose theirs in a short period of time provided the vapor tension of the surrounding gas is less than the vapor pressure of the crystals. As the crystal approaches the transition temperature of the hydrate, the loss of water from the crystal may become very rapid. Some dehydrations are so reproducible as to be used as secondary temperature standards. The dehydration of $Na_2SO_4 \cdot 10H_2O$ at 32.4°C is a well-known example. It is not so obvious that if the transition of $Na_2SO_4 \cdot 10H_2O$ is to be equilibrium reliable, it is necessary that the next less stable hydrated phase be in communication with the hydrate. There are three phases and two components. This means that the system has zero degrees of freedom if either temperature or pressure is held constant, as when the system is dehydrated in a vessel that is open to the atmosphere. At the transition there are two solid phases and the vapor. If the pressure is constant, then the temperature must remain constant until one of the two solid phases is lost. It is the same behavior that occurs during the melting of a pure component, at constant pressure, when the solid and the melt phase are the only phases under consideration. (See Chapter IV.) Furthermore, sodium sulfate forms a seven hydrate, $Na_2SO_4 \cdot 7H_2O$. In view of the statement of the succession of hydrates, one should logically expect the $Na_2SO_4 \cdot 10H_2O$ crystals to dehydrate to $Na_2SO_4 \cdot 7H_2O$ rather than Na_2SO_4. Phase diagram work has shown that $Na_2SO_4 \cdot 7H_2O$ is metastable under all conditions at atmospheric pressure. It can form only when there are no nuclei of anhydrous Na_2SO_4 to seed the system.

In dealing with the caking of crystals that form hydrates, there are several properties of the crystals that are paramount if the systems are to be understood and controlled. Crystal quality can usually be judged by how nearly perfect a crystal has been grown. Usually the more slowly a crystal is grown, the more nearly perfect it will be. This means that much attention should be devoted to the function of a crystallizer when a process is designed if a quality product is required. Faraday demonstrated that perfectly grown crystals of $Na_2SO_4 \cdot 10H_2O$ can exist for long periods under temperatures and vapor pressures that should have destroyed them. The more prefect the crystal, the more resistant it will be when subjected to conditions that would cause poorly formed crystals to cake.

Faraday performed numerous important experiments with $Na_2SO_4 \cdot 10H_2O$ that are necessary to the understanding of caking. He demonstrated that when nearly perfect crystals of $Na_2SO_4 \cdot 10H_2O$ were placed in a desiccator that they would remain transparent indefinitely but if the surface were scratched or the crystal damaged, the crystals would become dehydrated very quickly. Previously it was mentioned that the transition of $Na_2SO_4 \cdot 10H_2O$ to Na_2SO_4 anhydrous is so reliable that it can be used as a secondary temperature standard of 32.4°C, but Faraday also demonstrated that nearly perfectly grown crystals of $Na_2SO_4 \cdot 10H_2O$ could be heated to temperatures over 100°C without decomposing but if they were sprinkled with the smallest quantity of anhydrous seeds of Na_2SO_4, they decomposed almost explosively. In using the $Na_2SO_4 \cdot 10H_2O$ as a temperature standard, it is always mixed with Na_2SO_4 to seed the transition. This is a vivid illustration of the need for the presence of solid phase in equilibrium with the solution phase when solubilities are measured or aqueous phase diagrams are determined. The system under study can contain enough energy to be many calories away from a free energy of zero.

When considering the free energy and equilibrium of a phase transition, the free energy of the phase chemistry may be zero while the chemical free energy of the system may be very far from zero. Ammonium nitrate is a good example. The crystal phases at 25°C may be in equilibrium while the free energy of the following reaction may be very large. (−89.2 kcal)

$$2NH_4NO_3 \Rightarrow 2N_2 + 4H_2O + O_2 \qquad [5]$$

R. D. Young suggested the equation as a simplification of the detonation reaction of ammonium nitrate [32]. There are a large number of reactions occurring simultaneously during detonation but the standard free energy is probably of the correct order when the free energies of all the separate reactions are summed.

Repeatedly, several points based upon Faraday's observations have been discussed. Firstly a product should have a phase free energy near zero when it is packaged or loaded into cars for shipment. This means it should be in crystal equilibrium, but not necessarily chemical equilibrium. Secondly, the more perfect the crystals are, the more resistant to change they can be when subjected to conditions outside of their range of stability. Thirdly, the purer a system is, the more likely it is to withstand changes. Any impurity may act as a seed crystal causing a transition to occur that might otherwise be resistant to change. The action of seed crystals is poorly understood. Seed crystals can have the butterfly wing influence of a latent image of a photographic plate as mentioned in connection with chaos. The butterfly wing influence promotes the idea that if a butterfly flaps its wings anywhere on earth all other systems will be influenced to some degree.

An introduction to the two-phase, two-component constant pressure phase system is needed before continuing. Except for compositions that correspond to invariant points, eutectics, and pure compounds, mixtures of components do not melt at a constant temperature. It will be noted that when systems of this type are measured by differential thermal analyses, there is no melting point but a melting range. This system will be considered again when phase diagrams are discussed because it is fundamental when these diagrams are determined experimentally. Too often melting points are reported for systems which do not have a melting point because the compound is impure. It is recommended that some form of differential thermal analysis be used to determine melting points rather than the visual methods that are sometimes used. Much useful information may be lost with even hot stage microscopes if they are not supplemented with thermal analyses.

Consider the transition of a hydrate when temperature is held constant and the pressure is slowly changed (a three-phase system). Under this condition the pressure, *at equilibrium*, will remain constant until one of the solid phases is lost. Whether it is the temperature or the pressure that is held constant, the system will remain invariant as long as no phases are created or destroyed.

In order to review, consider what is happening when an anhydrous salt is hydrated. If water is allow to diffuse into the system as a vapor and the vapor is admitted very slowly to prevent any liquid phase from forming, it is possible, in theory, to form a hydrate without forming a saturated solution. Should one expect that the salt should cake? Very definitely. If the system is completely hydrated, not one single crystal of the anhydrous system still exists. All particles have changed shape and the bulk density of the system has altered. If the salt is contained in a bag, the bag has swelled. If the salt is contained in a drum, enough pressure can be exerted on the walls of the drum to cause them to deform. If the salt is now dehydrated in the same container, another set of anhydrous crystals will be formed. In most systems the new anhydrous crystals will be much smaller than the original, anhydrous crystals and the bulk density can be even less than the hydrated salt. The major point to remember is that phase transitions will normally cause systems to cake will cause lower bulk densities to be exhibited and will be of smaller crystallite size than the initial salt. Only one lump may remain but the crystals that form the lump are usually smaller than the starting salt.

Before leaving the hydration of an anhydrous crystal, it may be helpful to review what is happening with respect to the surface of the crystals and the molecules of water and the energy of the system. If a molecule of water is in the vapor state as a dissolved gas, it "contains" the heat of vaporization of a water molecule. If it collides with the surface of an anhydrous salt and sticks, it not only releases the energy of vaporization but also the heat of hydration of the crystal. This will create a local hot spot on the crystal. The energy that is released may drive the water molecule back off the crystal or if it remains on the surface, the energy release may keep other molecules from sticking at that point until the energy is dissipated. If the vapor tension in the atmosphere is greater than the vapor pressure of water in the hydrate, ultimately the crystal will be completely hydrated.

While considering the absorption of water by a hydrate, it is best to contrast what happens when a soluble salt that does not form hydrates or a soluble salt that is in its highest hydrated form at the specified temperature absorbs water from the atmosphere. The first thing that is different is only one molecule of water must be absorbed before a saturated solution behavior begins to exist on the surface of the particle. The word behavior was chosen because one molecule of water does not a solution make. Many molecules of water will be required before enough has collected on the surface to free a solute molecule or ion from the surface of the crystal. The second thing that is likely to happen is the temperature of the surface of the salt decreases, provided the heat of solution is more negative than the heat of condensation is positive. The solubility of the salt is an important consideration. The more soluble the salt, the less water is required to dissolve it and the less heat of condensation is liberated. Ammonium nitrate is an example of a salt that gets very cold when it absorbs water. The cooling effect causes the surface temperature to drop below the dew point and even more water is absorbed. This type of absorption does not cause caking, except for surface tension influences, but if the water evaporates from the surface, then crystal bridges form between the particles and the entire system can become one continuously bridged particle. (See Figures 2 and 3.)

Two other very important properties of crystalline hydrates are water capacity and the range of temperature stability of the hydrates. Alums are examples of systems that can retain twelve or more water molecules per formula weight of the anhydrous salt while having rather low solubilities. The second important property is demonstrated by the tenacity with which $Na_5P_3O_{10} \cdot 6H_2O$ clings to water. Under most conditions the tripolyphosphate anion will be destroyed rather then be dehydrated to the anhydrous salt. This is one of the primary reasons that $Na_5P_3O_{10}$ was the preferred detergent builder over sodium pyrophosphate, which loses its water of hydration rather easily. In Figure 4 a mixture of $Na_4P_2O_7 \cdot 10H_2O$ and $Na_5P_3O_{10} \cdot 6H_2O$ was heated at a constant temperature of 100°C and the thermogravimetric analysis curve clearly showed that the water of hydration of $Na_4P_2O_7 \cdot 10H_2O$ was much more easily lost than the water from $Na_5P_3O_{10} \cdot 6H_2O$. In Figure 5 sodium carbonate monohydrate was added to the mixture. This time the weight loss was measured as a function of advancing temperature. Although the sodium carbonate retains water more readily than sodium pyrophosphate, neither approach sodium tripolyphosphate as a water sink.

When working with the highest hydrates of a salt or an anhydrous salt that does not form hydrates, water can stack up on the

FIGURE 4. *The thermogravimetric analysis of two hydrates heated at a constant temperature of 100°C.*

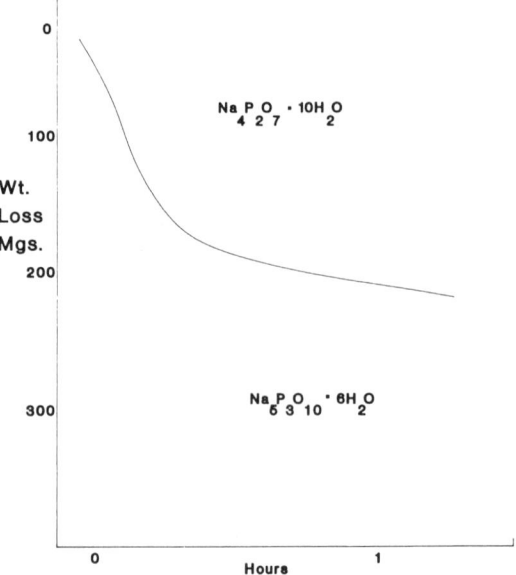

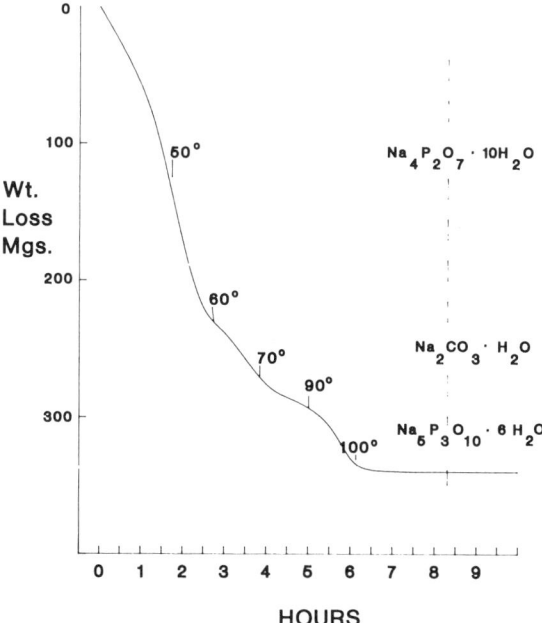

FIGURE 5. A thermogravimetric analysis of a mixture of three hydrates heated at advancing temperatures.

surface of particles more quickly than it can be spread over the surface of the particles. If the particles are formed from many small crystallites that have space between the crystallites, it is often possible to improve the behavior of the crystals toward caking by powdering a very small amount of surface active agent on the surface of the crystals. Sodium alkyl benzene sulfonate is particularly active in causing the water, otherwise trapped on the surface of the particle, to be distributed throughout the particle. If the system should then dry out, the salts dissolved in the saturated solutions will grow within the particles and not between the particles.

Hydrates have been considered mostly from an energetics and thermochemistry point of view. There are equally valid methods of viewing hydrates from a structural view. There is an excellent review of the structure of water in crystalline hydrates [33]. The structural science does not relate directly to caking but it does help to explain how water is held in solids and why some of the water is so loosely held while other water molecules are tenaciously retained in the lattice of the host.

In summary, any time a hydrate forms or is decomposed a phase transition occurs. In general phase transitions deform the particles of a product and the change of form is usually sufficient to cause the particles to bridge and form cakes. The formation of hydrates under some circumstances can be beneficial. Hydrates give holding capacity to some systems. This is particularly true in mixed systems where the primary substance is very soluble and does not form hydrates. Hydrate water that can move from phase to phase as a product is temperature cycled almost always results in a caked product. Differential thermal analyses, thermogravimetric analyses, and a temperature cycling chamber can do much to illustrate the behavior of these systems.

AMORPHOUS SOLIDS

Amorphous solids are difficult to define. It is much easier to state what they are not than it is to define what they are. Glasses are definitely amorphous solids and can be considered to be super-cooled liquids. But a crystalline solid can be ground so fine as to exhibit no x-ray pattern and will appear to be either glassy or cubic when viewed with the polarizing microscope. The work of Motooka and Kobayashi showed that it was possible to convert a crystalline solid to an amorphous solid by grinding the crystalline solid [34]. They showed that there was not only a change in molecular structure but also there was a change in the heat of solution of the salt after grinding. As pointed out in another section of this book, a definitive answer can be obtained from a differential thermal analysis of a salt before and after grinding. An amorphous solid will exhibit an exothermic reaction when heated if it crystallizes upon heating. If it does not crystallize, it will exhibit a glass transition, which is easily recognized. Most substances that can be converted to amorphous solids by grinding will crystallize again when heated.

The structure of glasses has been an interesting topic. Most of the interest has centered around silicate or phosphate glasses where the anions in the glass can exhibit molecular structure despite the fact that the total unit does not exhibit structure in the sense that a crystal does. It should be remarkable indeed if all of the molecules in a silicate or phosphate glass were exactly the same while the glass had no overall structure. Usually one can expect that these randomly formed systems should contain all manner of molecules and Van Wazer has shown that the phosphate glasses are a distribution of molecular sizes that depend upon the ratio of metal oxide to phosphorus pentoxide for the generally related structures [35]. The silicate glasses are probably more similar to the phosphates than is generally noted but they are more difficult to study.

The detailed structure of amorphous substances or even a definitive definition is not necessary in a caking study other than to note how they are likely to influence caking. In addition to the thermodynamic considerations of water absorption, amorphous solids present a special case in concretus science. They can be very difficult to deal with because they do not have a well-defined thermodynamic state, and therefore, cannot be considered to be at equilibrium under any conditions. Additionally, they do not have a solubility in the strictest sense of the word. This does not mean that the amorphous substance cannot be dissolved in water, but there is no fixed limit where the solution is in saturated equilibrium with a solid phase of defined properties.

Amorphous solids are unpredictable in other ways. It can be demonstrated that the electron environment about an SiOSi or POP linkage depends upon the metal oxide to anionic ratio. It can also be shown that the fewer electrons associated with a linkage, the more susceptible to reaction the linkage becomes to attack by metal oxides or water [36]. This also accounts for the "high energy bond" of adenosine triphosphate. Amorphous phosphorus pentoxide contains phosphorus atoms that are linked to other phosphorus atoms as triply linked phosphorus atoms. This means that a phosphorus atom is linked to three other phosphorus atoms by oxygen bridges. Triply linked phosphate structures are more highly reactive and explain why phosphorus pentoxide is an excellent desiccating solid. Specific reactions can contribute to caking in addition to the fact that the solid is amorphous.

The sodium polyphosphate glasses are excellent examples of the complications that can be encountered. Glasses always contain more energy than their corresponding crystals. Some glasses crystalize very rapidly while others are impossible to crystallize. If an amorphous solid crystallizes, it has undergone a phase transition and will have caked most of the time even if the crystallization is not complete. Amorphous solids have softening points rather than melting points, and a glass may deform and flow as a result of the heat of crystallization. This behavior can result in plastic-flow caking as well as embedded crystals caking.

Much attention has been given to glass transitions points in recent years. The glass transition is not a transition at all in the sense discussed here. There is no new phase formed at a glass transition point and a glass is a super-cooled liquid, the liquid merely becomes mobile.

Some phosphate glasses, as Graham's salt, are described as being infinitely soluble. The sodium polyphosphate glass shown in Figure 6 is transparent and is defined as infinitely soluble. As men-

FIGURE 6. *A water soluble sodium polyphosphate glass, "Sodium Hexametaphosphate".*

tioned, the description is inaccurate from the start. Infinitely soluble describes the property of the glass to form a one-phase system, all the way from the anhydrous glass to an infinitely dilute solution. Therefore, it is possible for a glass to absorb some water without caking, provided the water is absorbed slowly enough. If water is absorbed quickly, the surface of the particles become tacky and caking is certain to result.

Much work has been done over the years to control the rate of solution of sodium polyphosphate glasses. One approach has been to add multivalent cations to the sodium phosphates while keeping the M_2O/P_2O_5 ratio near unity. One commercial water treatment glass sold for home use is a mixed oxide with roughly the composition $CaO-Na_2O-P_2O_5$. It dissolves slowly enough to allow the water to flow through a bed of the glass and is very effective in suppressing red water, caused by iron, and green water, caused by copper. Other ways to lower the rate of solution is to modify the anion with other anions that react with the polyphosphates. It has been noted that when the rate of solution is lowered the rate of absorption of atmospheric water is decreased and the caking tendency of the glass is also greatly suppressed.

When using differential thermal analyses, very often systems are more likely to superheat than to supercool. Heating curves are obtained when a sample is heated from a lower temperature to a higher temperature while recording the differential temperature of the sample with respect to some standard substance, such as aluminum oxide, as a function of the temperature of the sample. It is usually better not to record time as an explicit variable as is done with some commercial instruments. Details of the thermocouple design and instrumentation shall be discussed in the section dealing with tests.

When working with glass formers, such as phosphates and silicates, there is no option; heating curves must be used because very long periods of time may be required for the sample to crystalize. If excessively long times are required, the rate of liberation of energy will be too slow to be accurately detected by most instruments. In working with *heating curves* it can be shown that all stable phase to stable phase transitions are endothermic and that if an exothermic reaction occurs, either a chemical reaction produced heat or a metastable substance has converted to a stable substance. This is easily seen from the well-known relation between free energy, G, enthalpy, H, entropy, S, and temperature, T.

$$dG = dH - TdS \qquad [6]$$

At a transition point, dG is equal to zero because at the transition temperature, two crystalline forms coexist and if the conditions are handled correctly they can be in equilibrium. At equilibrium the driving potential for a change is zero; dF = 0. Therefore

$$dH = TdS \qquad [7]$$

As energy is added at the transition, dH is positive while T remains constant until the conversion from the low temperature form to the high temperature form is completed. The higher temperature form will contain more energy than the lower temperature form. The transition from a stable low temperature form to a stable high temperature form is endothermic. This is exactly the same reasoning applied to the ice point where ice and water are in equilibrium at 0°C. When heat is added to the system, ice melts and when heat is extracted from the system, water crystallizes but when carefully done, very slowly, the temperature remains 0°C despite that heat is either being added or extracted. The system will remain at 0°C until all of the ice melts or all of the water freezes. Either way, the systems moves from a two-phase system to a one-phase system while pressure is held constant. Temperature is not held constant but when two phases are in equilibrium in a one component system at constant pressure, the phase rule is:

$$F = C - P + 1 = 1 - 2 + 1 = 0.$$

Where:
  F is the degrees of freedom.
  C is the number of components.
  P is the number of phases.

Consequently, the system has zero degrees of freedom and the temperature remains constant just as long as two phase, ice and water, or crystal I and crystal II coexist. For a more detailed discussion of the phase rule applications, see Chapter IV.

PHYSICAL PROPERTY MEASUREMENTS

There are many ways to approach the solution to a problem. Very often there is no best way. It is always true that one fact is worth more than a book of speculations. In order to apply linear differential equations to describe a system, authors are inclined to utilize models that are oversimplifications of the true physical system. They are usually pleased when the derived mathematics are in reasonably good agreement with measured systems. It is possible in these cases to obtain speculations disguised as facts. These models may be useful but one is warned to test the model well for the system under consideration.

In Chapter VI the author includes many of the tests for physical properties that have worked well for him. The tests as well as the research projects have followed a common theme, "Keep it simple". Tests for tests sake can do more to confuse and confound than to enlighten.

SOLUBILITY

One of the prime considerations in many caking problems is the solubility of the components in the caked system. Unfortunately most theories of solubility have not been of value to the practicing scientists. If the solubility of a given system is desired under a new, untested set of circumstances, it is best to measure the solubility of the system because even the best theories are likely to produce poor estimates. Under these conditions an incorrect number can be worse than no number, particularly if a product's future is depending on the validity of the number.

There are a number of simple rules that should be recognized before continuing with a study of pure solids dissolving in pure liquids. Pure solids and pure liquids are chosen merely to simplify the discussion and to better relate to the spontaneous caking problems.

A definition of solubility, with respect to *solids,* is required before continuing to discuss the subject. Liquid solubility in liquids and gas solubility in liquids are not considered here. When referring to solubility the reference will be to the quantity of a thermodynamically stable crystalline substance dissolved in a defined quantity of a thermodynamically stable solvent at a fixed pressure and temperature at saturated equilibrium. To establish equilibrium, metastable phases are prohibited by definition. Time is required for equilibrium to be established. How much time will depend upon the system under consideration and the physical, chemical, and mechanical factors contributing to the approach to equilibrium. Equilibrium will be established when the rate of solution of the solid phase, dS_s/dt, is equal to the rate of crystallization of the solid from the liquid phase, dS_l/dt.

The factors contributing to the rate of attainment of equilibrium are many. The ultimate concentration of the solution at equilibrium, the particle size of the solid phase, the temperature of the system, the mechanical agitation of the system, the volume ratio of solid phase to the liquid phase in the system when equilibrium is established, the tendency of the solid phase to become supersaturated, the rate of crystallization from supersaturation, hydrate formation in the solid phase, and the rate of hydrate formation. Other variables may come to mind but these are several of the more important ones. Great care must be exercised if solubility measurements are to be useful. This is particularly true if an approach is followed where excess solids are added to a system and equilibrium is established when two solids have reacted to yield either a double salt or a solid solution. If possible it is better to allow the systems to become homogeneous by heating the solutions to a temperature above the measuring temperature. If this is not possible, then systems may be prepared that have very little solid phase compared to the liquid in the system.

GENERAL RULES

1. Solubility is an equilibrium concept, which is defined by the laws of thermodynamics and consequently the phase rule.
2. For solid systems, only crystalline solids have a true solubility.
3. Many amorphous substances dissolve but there is no reference state that is defined because an amorphous solid phase is itself metastable with respect to some crystalline phase, whether or not the amorphous phase has yet been crystallized.
4. Not all crystals that dissolve have a solubility. If the crystal is not stable but can react with the solvent, for example, then the crystal has no defined solubility. Hydrolytic degradation is one example of this type of instability. Disproportionation is another type of degradation of a solid phase that prevents equilibrium from being established with respect to the original solid phase. An equilibrium may be established with respect to the degradation products but in this case it is incorrect that the system may be considered as only two components.

Much work has been done in an attempt to establish the solubility product of hydroxyl apatite, approximately $Ca_5(PO_4)_3OH$, but most of the values are not very reliable because the surface of the apatite is constantly changing as a result of hydrolysis [37]. The interest is great because hydroxyl apatite is so important in the formation of bones and teeth [38]. The use period may extend over many decades. Reliable values for the rates of dissolution and the quantity dissolved could be very useful information. At best, approximations must be accepted because hydroxyl apatite shall forever be thermodynamically unstable in aqueous media.

5. Solubility, as normally measured, is determined with pure substances at atmospheric pressure and a specified temperature. It is normally a two-phase, two-component system. This means that normally, one pure crystalline solid is in equilibrium with a solution of the pure solid dissolved in one pure liquid.
6. If all of the crystals dissolve and only a solution phase is left, the solubility of the crystals in the liquid cannot be determined until the solid phase is present by either adding more crystals or by evaporating enough of the solvent for crystals to form. It may be possible to obtain crystals by cooling the solution, but they will again dissolve as the temperature is returned to the measuring temperature. In order for the system to be at equilibrium, and therefore invariant, it must have zero degrees of freedom. For this to happen at constant temperature and pressure, the number of phases must equal the number of components. There are two components in the system. Therefore, there must be two phases, ignoring the vapor phase. For equilibrium to be established, some solid phase must be present, in contact with the solution at saturation. The vapor phase may be ignored only because the solubility is measured at atmospheric pressure and this is assumed to remain constant during the course of the experiments. The atmospheric pressure seldom remains constant for this time but the influence is so small as to be safely ignored when determining the solubility of a solid in a liquid at constant temperature. Obtaining a constant temperature is usually much more important and more difficult to control within the required limits.

When the solution phase is in contact with the solid phase stable under the chosen conditions, a reference is established. It may be thought of as a thermodynamic anchor point for the system. The quantity of solid phase is not important because the equilibrium is dependent upon intensive variables. At constant temperature and pressure the phase rule is:

$$F = C - P \qquad [8]$$

where:
F is the degrees of freedom
C is the number of components
P is the number of phases.

If F is zero (invariant), then C must equal P.

THE PHASE RULE

To illustrate the need for a stable solid phase while establishing an equilibrium solution, sodium tripolyphosphate is not thermodynamically stable in the presence of water. It undergoes hydrolytic degradation. In a strict sense of the concept, sodium tripolyphosphate can have no solubility. The solution phase is never in equilibrium with the solid phase because the solid phase is constantly changing.

$$Na_5P_3O_{10} + H_2O \Rightarrow Na_2HPO_4 + Na_3HP_2O_7 \qquad [9]$$

The reaction may be ever so slow, but the system will be constantly changing nevertheless. The author has published aqueous phase diagrams of condensed phosphates but it was always recognized that the systems were at best close approximations of the systems had it been possible to establish equilibrium.

The so called, Sodium Hexametaphosphate glass, of commerce cannot have a solubility for two reasons. Firstly, it is amorphous and secondly, it is hydrolytically unstable in the presence of water.

When the solid Sodium Hexametaphosphate is heated dry:

$$Na_2PO_3 \left[NaPO_3 \right]_{n-2} PO_4Na_2 \Rightarrow (n-2)/3\, Na_3P_3O_9 + Na_4P_2O_7 \qquad [10]$$

(amorphous) \qquad (crystalline) \qquad (crystalline)

And in solution:

$$Na_2PO_3 \left[NaPO_3 \right]_{n-2} PO_4Na_2 + H_2O \Rightarrow Na_2PO_3 \left[NaPO_3 \right]_{n-2} OH + Na_2HPO_4 \qquad [11]$$

or also in solution

$$Na_2PO_3 \left[NaPO_3 \right]_{n-2} PO_4Na_2 \Rightarrow Na_2PO_3 \left[NaPO_3 \right]_{n-5} PO_4Na_2 + Na_3P_3O_9 \qquad [12]$$

Both Na_2HPO_4 and $Na_3P_3O_9$ are crystalline salts when the water is driven from the solution.

The average n used in the above equations is near 8 to 10 in the Hexametaphosphate glass of commerce. The n is the average chain length of the polyphosphates in the glass. In the sense n has been used, it is the total number of phosphorus atoms in a single polyphosphate molecule and is usually reported as the average number of phosphorus atoms per molecule because all phosphate glasses are composed of an array of sizes of molecules. In preparing polyphosphate glasses the mean of the distribution is merely shifted higher or lower depending upon the property desired.

HEATS OF SOLUTION

Several other aspects of solutions and solubility are worthy of consideration. The heat of solution of a substance is often presented as an exothermic quantity. To avoid the problem of the reversed signs of heat of solution and enthalpy of a system, exothermic and endothermic will be used to indicate whether heat is leaving a system under consideration or the system is absorbing heat from its surroundings. (A positive heat of solution means an exothermic reaction while a positive enthalpy of reaction means that an endothermic reaction has occurred.) *The heat of solution of a stable salt is never exothermic!* All salts which do not form solvates (hydrates) and all salts in their highest solvated (crystalline hydrated) form, at the chosen temperature, dissolve endothermically! It is similar to discussing an exothermic heat of vaporization, to discuss an exothermic heat of solution. Heats of solvation (or hydration) are exothermic and they should always be separated from heats of solution. Unfortunately, this is seldom done in the older literature.

To this point in this discussion the ideas of partial molal quantities, differential heats of solution, and integral heats of solution have been intentionally ignored. Just for the record, the discussion centers about the integral heat of solution from pure solvent to saturation. The reader desiring additional information can become thoroughly confused on the issue by reading one of the standard texts on chemical thermodynamics. It becomes obvious that many of the authors haven't cleaned a beaker in the twenty years prior to writing a description of heats of solution.

Before leaving heats of solution, perhaps it will be helpful to mention why hydrate formation is exothermic. Hydration is exothermic because the water molecule that becomes a hydrated water molecule, lost most of its kinetic energy that it had been using in roaming around a beaker, when the water molecule is stopped, dead still, in a crystal lattice. The kinetic energy leaves the system as heat of hydration. On the other hand, as a salt is dissolved it does just the reverse.

The ions or molecules in a crystal lattice have no kinetic energy of the roaming kind.

If the salt does not form a hydrate or if it is in its highest hydrated form at the temperature of the system, the entire system increases in kinetic energy as the salt dissolves. The increase in kinetic energy is supplied from the surroundings. The solution first cools and then absorbs energy to return to the original temperature of the surroundings. It is the energy absorbed from the surroundings that is mostly converted into the kinetic energy of the system. If heat is evolved when a salt is dissolved, the salt is either not in its highest solvated form, it is metastable with respect to some other crystalline form, or a chemical reaction, hydrolysis, neutralization, or the like has occurred.

If one is to understand what is happening as a substance absorbs water from the atmosphere to cause the caking of a solid, it is necessary to understand the simple but rather involved explanation given above. It will be referred to and restated in many forms, throughout this work. It is at the very heart of what is happening in more than half of the caking problems likely to be encountered.

In a quiescent system solubilization is quickly spontaneous only on the surface of a crystal. When a layer of saturated solution exists on the surface of a crystal, the crystal is only very slowly dissolved and recrystallized. The equilibrium is dynamic. If more salt is to be dissolved, the solution on the crystal must migrate to expose fresh surfaces or the solution must be agitated, thus diluting the saturated solution on the surface of a crystal. Diffusions in liquids are a very slow process. Years may be required for even a small system to reach equilibrium if only diffusion is allowed to mix a system. But when a soluble salt is exposed to the atmosphere, even without stirring, several factors may cause the salt to continue to absorb water even after the surface is coated with a saturated solution. Water from the atmosphere is continually diluting the surface water at the surface and agitating the surface with the heat of condensation of the water molecules. Unlike the crystals contained under water on the bottom of the container, the solution phase may migrate by gravity exposing new surfaces. Consider a crystal suspended in a solvent. The crystal continues to dissolve because the density of a solution streaming from a crystal is of greater density than the partially saturated solvent. At saturation, the dissolution of even the suspended crystal ceases although the dynamic nature of the system will become evident as the shape of the suspended crystal continues to change, usually toward a more perfect crystal.

It is an interesting sideline that most solid or liquid atomic elements are not soluble in water, at atmospheric pressure, and room temperature. The elements that react with water are usually soluble, but not as the element. Also, salts with high densities are usually less soluble than lighter crystals, but there are many exceptions. The salts of the heavy metals are usually less soluble than the corresponding salts of the lighter elements, but again there are many exceptions. It is unfortunate so little progress has been made since the Greeks pondered the question of why things dissolve. The solubility of a salt in water cannot be predicted and the expression "saturated solution" is meaningless. What fundamental property of the system is saturated? It is unfortunate that measurements are still required to know the solubility of a salt in water if the salt's solubility in water has never been measured.

MIXED SYSTEMS

To this point the dissolution of a pure salt in a pure liquid has been discussed. If the solvent is a mixture of a liquid and two dissolved salts things begin to get a little more interesting. Perhaps only a solid phase and a liquid phase exist in the system, but the system contains three or more components. The system, which was invariant at a fixed temperature and pressure, now has one or more degrees of freedom. Something more than temperature and pressure must be specified if it is to be known where the equilibrium conditions are for the system. It should be recognized that when one speaks of knowing where the equilibrium conditions are, the phase diagram for the system has already been determined and the specification is where the equilibrium is established on the particular phase diagram in question. The process is similar to knowing a location on a map when the coordinates are arbitrarily specified as they often are on road maps. If the location is in Mexico and the coordinates are applied to a map of China they mean nothing.

In the previous discussion, composition is more than just an intensive variable. It has some extensive properties. How much of the components, relative to each other, has meaning. For the sake of illustration, assume that a second salt has been added to one of the solutions of a pure solid dissolved in a pure liquid already discussed. The system of two components was saturated and is at equilibrium. What has happened when the second salt is added? There is at least one more component in the system. This assumes that the added salt may be treated as a single component. When the second crystals are added to the saturated solution, the solution is no longer saturated and is no longer at equilibrium. It must establish a new equilibrium but where has not been determined. If only a small quantity of the third component is added and it all goes into solution, the new equilibrium could be established almost any place. This depends upon the nature of the new system. But if enough of the third component (extensive) is added to the system and it does not all dissolve with some of the first equilibrium solid remains and the two solids do not react with each other to form either a double salt or a solid solution, there are again as many components as phases and the system is again invariant. Only one solution of a very precisely fixed composition can be in equilibrium with the two solid phases simultaneously.

In considering the addition of a second solid substance to a two-component saturated solution in which equilibrium had been attained, the complexity of the system will depend upon the number of components in the perturbed solution. If the original solution was composed of water and a crystalline solid composed of neutral molecules and a second substance composed of crystalline neutral molecules was added to the solution at equilibrium, the number of components should be three.

If, on the other hand, the original two-component solution was composed of an ionized salt and water and when the second salt was added to the solution, the system contained only three components this will usually mean that the added salt has an ion in common with the dissolved salt. To be certain that this point is understood it will be belabored a bit. Assume that the original solution was composed of water and sodium chloride. It can be treated as a two-component system because there is an equation

$$2Na + Cl_2 \Rightarrow 2NaCl \qquad [13]$$

relating the reactants to the product for the NaCl and a second equation

$$2H_2 + O_2 \Rightarrow 2H_2O \qquad [14]$$

for the water. There are four elements but the two equations reduce the components to two. In working with the phase rule it is recognized that the same logic exists as is found in solving simultaneous equations. If one is to solve a system of simultaneous equations for all unknowns independently, it is necessary to have as many independent equations as there are unknowns. The phase rule merely states the number of variables, $C + 2$, minus the number of phases, independent states. If independent equations exist between some of the variables, their numbers may be reduced. The points are clearly seen when the phase rule is derived as Gibbs derived it based upon chemical potentials.

These and other systems will be reinvestigated in Chapter IV, dealing with phase diagrams. They are a natural outgrowth of the solubility concept, but they become more interesting and more useful when dealing with caking problems.

GRINDING

Grinding can have both positive and negative influences on the caking of solids. Usually, the finer a powder is ground the more likely it is to cake. While discussing amorphous solids above we noted that it was possible to convert crystalline solids to amorphous solids by exteme grinding. Grinding can also grossly influence the flow properties of a solid. This can happen in a variety of ways. One of the less obvious is the formation of fibers from an otherwise granular product by cleaving the crystals along an axis that yields fibers. The diamond shaped free flowing powder of $[NaCa(PO_3)_3]_n$ is converted to phosphate fibers which flow very poorly, after grinding. (39) The same is true of many other natural and synthetic products which cleave longitudinally when impacted with sufficient vigor to destroy the original crystals.

A second flow and caking problem that may occur as a result of grinding is the absorption of water. There may be some amorphous phase formed, which is usually more hygroscopic than crystalline phases. The temperature of the product is likely to be sharply increased during impact. This will not only cause the sample to absorb water more readily as the product cools but any hydrates in the product may be completely or partially destroyed. It is even possible that solid crystal-solid crystal phase transitions may occur during grinding. Additionally if the product is either piezoelectric, pyroelectric, or even ferroelectric, the forced ionic dipoles created in the crystals by both the formation of strain or the release of strain can contribute to caking and flow problems. If these were not enough, static electricity may become very noticeable during and after grinding, particularly in very dry, cold weather.

There are many other factors that should be considered when it is necessary to grind a product before it is shipped or stored. The main point to consider is that grinding may cause very subtle problems and that there are many types of mills that should be considered when milling is found to be at fault. It is possible to purchase mills that are both cooled and desiccated by dehumidifying the air dragged into the mill while it is in operation. Fortunately most milling problems can be solved, but there may be difficulties in isolating a problem to a mill if one does not anticipate milling as a problem area.

FORMULATIONS

Solid formulations are a new dimension in the type of systems being considered. Although phase chemistry is involved in solid formulations, it is not typical of the phase chemistry that has been considered to this point. In this case one must consider reactions of the type where an oxidizing agent and a reducing agent are mixed in the same container. This is the type of problem that may occur when a bleaching agent is mixed with certain types of surface active agents. It is the type of problem that has caused explosions of magnesium perchlorate, potassium nitrate, and ammonium nitrate that have been mixed with organic compounds. Even if explosions do not occur, the value of the oxidizing agent may be diminished and the phase chemistry predicts that the system is very likely to cake. Other examples are acids and bases contained in the same product. This is more likely to happen in the pharmaceutical industry, but it is not unknown in the animal feed, pet feed, and fertilizer industries. Calcium carbonate is found in many animal feeds because it is a safe nutrient and is inexpensive. Calcium carbonate is very reactive with acids and in the presence of acids can cause caking if it is not protected.

Ammonium salts of all types may also react with bases. Ammonium phosphates mixed with sodium carbonate will not only generate carbon dioxide, ammonia, and water but a caked product as well. Therefore, it is imperative to keep products capable of generating water dry at all times. Once the reaction is under way, it is self promoting generating more and more water. Water of composition can be converted to free water. It is similar to the reaction of a mixture of sodium carbonate and phosphorus pentoxide where heat, not water, is formed. As long as the system remains dry, nothing happens but when one drop of water is added the reaction is initiated.

$$2Na_2CO_3 + P_4O_{10} \Rightarrow Na_4P_4O_{12} + 2CO_2 \qquad [15]$$

Carbon dioxide is generated at once and the reacting mixture becomes very hot and may even form high temperature melts without being heated. There seems to be no limit to the quantity of product that could be generated in this way. Needless to say, under most conditions the reaction product is badly caked, but under conditions to form tetrasodium pyrophosphate the product does not cake badly and can even remain free flowing after the reaction is completed.

Yet another problem that can occur is the precipitation of an insoluble salt from two soluble salts. Formulations containing soluble calcium or magnesium salts such as chlorides, acetates, or nitrates, when mixed with sulfates, oxalates, fluorides, phosphates, or carbonates are very likely to cake badly by the formation of insoluble salts.

These are but a few of the hundreds of combinations that can prove to be incompatible by solid-solid reactions yielding new phases. Water is likely to trigger the reactions that would probably not occur in an anhydrous system. Also if any of the salts are hydrates, a source of water is built into the mixture and if the temperature rises above a transition temperature the newly formed, less soluble salt may or may not form hydrates.

As mentioned, too often rail cars or trucks rented or owned by the manufacturer and used to transport solids are not weatherproof and watertight. It may appear that the only problem caused by water leaking into a car is the localized lump that can be removed but it is very likely that the time required to unload the car is many times greater than if the car had been well sealed. The usual excuse for using a car that is not well sealed is based on the need to meet a production schedule and the car in hand is the only car available when it is needed. This is poor planing. With modern computers every car or truck has an identification code and once the data base is established, how it may be used has been mentioned as it relates to caking problems. Here is a case where the statistical treatment can quickly reduce the quantity of work required to yield THE optimum formulation from the ingredients required, particularly if some reasoning is applied to the system before the experiment is defined. There is no need to have a statistical treatment direct a project to facts that are known before the analysis was undertaken.

DETERGENTS

Few products are likely to be more closely scrutinized than household detergents. Because non-phosphate detergents are surely to be the dominant item of commerce in the future, special attention will be directed to these detergents. Much of what can be considered about detergents can apply to other products. There is much caking of detergents in the future and some of the new formulations in the old familiar boxes are beginning to cause caking problems. About the only thing left that is familiar is the old familiar box is the brand name on the box.

In order to produce a non-phosphate detergent that will perform, it is usually necessary to increase the pH of the wash water. This is usually done by formulating with more sodium silicate, sodium carbonate, or even free sodium hydroxide. Combine this with the fact that some kind of bleach is added to the detergent as well as possibly a bleach activator to help boost the cleaning power of the detergent and the systems become more and more complex. A builder, which may or may not be included, can be loosely referred to as a water softener if one is not a stickler for exactness. Two or more surfactants may be used to help performance in hard water if a builder is not used.

It has been mentioned that each new ingredient in a formulation vastly complicates the system from a phase chemistry point of view but also from a formulators point of view. It has been known for many years that the order of addition of the ingredients to a formulation can make much difference in the physical properties of the product. Assume for a moment that a system is formulated with the following raw materials:

1. Sodium carbonate
2. Sodium sulfate
3. Sodium silicate
4. A builder as sodium citrate, sodium tartrate, or NTA nitrilotriacetate.
5. Two or more surfactants
6. A bleach as a perborate, percarbonate, or peracid salt
7. A bleach activator
8. Perfume
9. Optical brightener
10. A hydrotrope

Assume that the formulation contains eleven different ingredients, and they are blended in a mixer or crutcher. There are eleven ways to choose the first ingredient. One is chosen. Toss it into the mixer. Now there are only 10 ways to choose the second ingredient but this means that there are 110 ways to choose the first two. As more and more of the ingredients are chosen the system gets more and more complex. There are eleven factorial (11!) different formulations merely by the different sequence in which they can be mixed. $11 \times 10 \times 9 \times 8 \times 7 \times 6 \times 5 \times 4 \times 3 \times 2 \times 1 = 39,916,800$.

The former may be considered only as an exercise. Be assured that it definitely is not! The order of addition will have much to do with whether or not free water, free sodium hydroxide, activated bleach, reduced shelf life, or caking will occur. Obviously, it is not practical to test all forty million mixtures. Two approaches may help. First, an understanding of the chemistry is mandatory. Second, a good statistical approach will be helpful but must not be expected to be a panacea.

After all of the previous, there may be no way to mix the eleven ingredients that will prevent caking if the box is left open on a hot, humid day. The following is an outline of how the problem may be approached. It is one of at least forty million ways the problem may be approached. The following is merely one of the approaches that is based on a logic that can be helpful if the caking problem has a practical solution.

Considerations for formulating:

No. 1. Prepare a list of the solubility in water of all of the ingredients to be used. If the solubility has not been reported in the literature, it must de determined according to the outline in Chapter VI. If the substance is similar to sodium silicate liquid and has no solubility, its behavior should be carefully noted. Improperly used sodium silicate liquid can lead to free water, free sodium hydroxide, or both. Free sodium hydroxide can lead to carbonization from the carbon dioxide of the atmosphere or hygroscopicity from the water in the atmosphere.

Once the solubility of the ingredients is known, the more soluble ingredients are suspect. Determine the rate of absorption of the more soluble salts as a function of relative humidity at room temperature (25°C to 30°C). If an ingredient absorbs water at a vapor-loading as low as 40% relative humidity at 25°C, a substitute should be found for the substance, if possible. It is unlikely that even encasing the grains of product would keep it from caking under humid conditions if it were extremely soluble.

The "bad apple syndrome" must forever be at the forefront of all works with formulations. Only one substance in a formulation array may absorb water from the atmosphere. Sooner or later the soluble component will cement the product into a brick despite the fact that it maybe only five percent or less of the formulation.

No. 2. List the number of substances in the formulation that are known to be crystalline. Include in the list the known hydrates and the ones that are known to be stable at ambient temperatures and humidities. If the decomposition temperature of the hydrate is known, note this temperature but in addition run a differential thermal analysis and a thermogravimetric analysis to determine the stability of the hydrate as a function of temperature. Any hydrate that is loosing hydrate water as low as 80°C is suspect. It is salts of this type that will allow water to escape from itself and end up hydrating another phase or even being lost to the atmosphere. Either event may cause caking to occur. The caking is likely to be more severe if the water is moving around in the container than if it is lost to the atmosphere because more of the system is undergoing phase changes.

When formulating if free water or that contained in a solution is added to a system, it is better to mix the water with a salt that hydrates water tightly than it is to mix the water with a substance that forms no hydrates. This is more common sense than either science or statistics. The more that is known about a system before statistics are applied the less work there will be to arrive at an acceptable answer.

No. 3. Similar considerations govern amorphous substances as govern crystalline solids. They should be listed and their tendencies to absorb water, carbon dioxide, or other components of air should be noted. If data is not available, it may be necessary to obtain hygroscopicity and similar data in the laboratory.

Silicate solutions are commonly used and are often converted to amorphous solids when the solvent is extracted or distributed in other components of a formulation. Consideration as to how these solids are added is important. If substances such as clay or zeolites are added to a formulation, it is unlikely that a silicate solution, which may turn to an amorphous solid, should be added directly to insoluble ingredients. This may either cement the particles together or seal active channels.

In some formulations the ingredients may themselves be formulations supplied by another firm. This may cause complications if ingredients have been compounded elsewhere. The more likely case is the ingredients in the formulation will be mixed on site lending control as to where and how they are introduced.

No. 4. Organic materials may be either solids or liquids, crystalline or amorphous and may be treated as a special case. The major problem that can be expected from the builders in the new non-phosphate detergents is their greater solubility and less stable hydrates. Unless some method can be found to control the solubility, as discussed in Chapter IV, these will continue to give troubles.

Some of the organic ingredients are composed of a number of optical isomers, each behaving as phase chemistry component, yielding a control nightmare. These ingredients are used to obtain formulations that will perform but are short lived after their initial function has been fulfilled.

Usually, properly applied, the active surfactants can

help a system to be free flowing. This factor is a definite part of the order of formulation and the capacity of the other ingredients to imbibe and hold liquids and waxy solids. Improperly applied the formulation can "bleed" an oily liquid on to a container or package. This will not normally cause caking but it can be as disagreeable, particularly if the oils bleed on to garments, furniture, upholstery and similar surface where it is not desired. Experimental designed, statistical studies should be considered for study of these formulations.

No. 5. In the search for new ingredients it is helpful to compare the ingredients with respect to the properties of sodium tripolyphosphate. Sodium tripolyphosphate has water holding capacity. It also holds the water very strongly. It has a relatively low solubility. It has a high imbibing capacity for organic actives. It is thermodynamically very stable in the box, reacting with neither atmospheric water or carbon dioxide. It is rapidly "biodegradable". It forms strong complexes with calcium and magnesium ions. It is an excellent buffer in the exact pH range to give excellent cleaning and without creating free sodium hydroxide. It is completely toxicologically safe. It is completely bleach stable and it has an infinite shelf life. It is inexpensive. It is the primary reason that the old style detergents do not cake.

Rather than searching for a single substance with similar properties to sodium tripolyphosphate it will be much wiser to search for systems of two or more ingredients. In this case the phase chemistry cannot be ignored. Single ingredients may have satisfactory properties while mixtures may be incompatible.

The foregoing discussion is intended to point out that if the caking problems that will occur with the new formulations are to be solved while maintaining performance and cost much fundamental research will be necessary. The hopes of merely statistically formulating a product out of trouble when there is a fundamental thermodynamic flaw in the product is wistful thinking. Only well trained scientists can be expect to have the knowledge base necessary to compete in a field whose tradition is fundamental science and continuous change. The problems will be difficult but possible to solve, and new and better methods of formulating detergents are certain to follow.

Summary / 77

THE FLOW OF SOLIDS

It has been tacitly assumed throughout the discussions of the chemistry of the caking of solids and the flow of these solids that in most cases the attempt was made to obtain solids that behaved as liquids when the solids were desired to be put into motion. It makes little difference whether or not a powder is caked if it is never to be moved in any way. When the solids are to be moved, it is desirable that they are not caked and that they flow as a liquid. One vivid example is the use of white phosphorus rather than red phosphorus in industrial applications of this element. White phosphorus melts at 44.1°C and can easily be pumped, transported, and metered into processes in its liquid state. Red phosphorus has some desirable physical properties that could make it the preferred form of phosphorus but it is a solid that has a high melting point and must be handled in equipment that is more difficult to used and processes must depend upon weight measurements rather than volume measurements.

The solids have an advantage in that their vapor pressures are usually very low compared to liquid. This is usually but not always true as can be seen by allowing ammonium chloride to sublime.

SUMMARY

In this chapter it has been demonstrated that any time particles start out as free-flowing, cake, there must be a driving force to cause them to adhere to each other. Particles cake to remove some kind of a stress either imposed on the system from external causes or from stresses contained within the particles themselves. In any event the system will be in a lower energy state after the particles cake than when they were a free-flowing powder. This, of course, considers that energy may be brought into the system as water vapor, carbon dioxide, or similar substances. It does mean that energy must be supplied to the system to break up the lumps and cakes if the particles are to be returned to their original free-flowing condition.

Gravity is one of the external forces that is always present. The influence of gravity will depend upon circumstances. If the particles are at the bottom of a stack or rail car the influence of gravity is much greater than when the particles are at the top where only their own mass is acted upon by gravity. Gravity is the most likely cause of the influence of piezoelectric action even when the particles are jostled in a freight car. Plastic flow can be a direct result of gravity. The ever present water vapor can be involved in the release of stress imposed by gravity and other causes. Water works with gravity to cause caking with the release of stress.

The migration of mass across crystal interfaces can be driven by piezoelectric action and aided by water vapor to increase the mobility of the mass at the interfaces. Water vapor is also capable of lowering the free energy of the system by forming hydrates, saturated solutions on the surface of solids, hydrolysis of compounds, and causing solid-solid phase transitions to occur at lower temperatures. Other reactants may be carbon dioxide, hydrogen sulfide, or ammonia, all of which can cause a system to cake.

In most cases formulations are more inclined to cake than pure substances because their chemistries are much more complex. Water may cause two components to react that might otherwise remain dormant without the influence of water. If the chemistry of a system is understood, caking may be eliminated in many products, which might be impossible to control without this knowledge.

The dominant role of water in the chemistry of solids caking is reinforced by the general chemistry of caking. Even in those cases where water is not the ultimate cause of caking, it usually plays a secondary part in the final results. Adsorption and the colloid science associated with adsorption is instrumental in causing a substance to cake. The adsorption may be either chemisorption or merely physical adsorption.

When hydrates become involved in the caking process, they may be either beneficial as inert water sinks or they may be harmful if they do not hold water tightly in the crystal lattice. Also of importance is the type of surfaces involved. As expected amorphous and crystalline materials do not behave the same and the differences may dictate whether a system cakes.

Solubility in water is perhaps the greatest determining factor that must be controlled if an item of commerce is to remain free flowing and highly soluble salts must be protected if they are to remain free flowing. No satisfactory fundamental theory of solubility exists and usually a system must be subjected to tests to determine how soluble it is in water. A few simple rules of thumb for solubility may be used but these are little more than what may probably be expected but may be helpful while executing laboratory measurements.

The mere grinding of solids in high impact mills may have a drastic influence on the tendency of a system to cake. If the crystals cleave to fibers, they can become entangled causing mechanical caking. Other factors influencing caking may be mixed systems such as those occurring in formulations. Formulations may become very complex particularly if the order of addition of the components is very important. Usually dry powders may be mixed in almost any order or miscible liquids may be mixed with each other without changing the properties of the system, but when liquids and solids are mixed entirely different products may result from mixing orders.

Chemistry is at the heart of most caking problems but it is not the only reason that solids cake. And no general factor is more important than the chemistry of water and how it interacts with other parts of a system.

REFERENCES

20. Harned, H.S. and Owen, B.B., *The Physical Chemistry of Electrolytic Solutions,* p. 513, Reinhold Publishing Company, New York, N. Y. (1958).
21. Glasstone, S., *Textbook Of Physical Chemistry,* p. 651, D. Van Nostrand Company, New York (1946).
22. Shpak, E.A., Chesha, I.I., Adamenko, V.V., Datsenko, D.F., and Verbetskaya, T.G., *Khimicheskaya Tekhnologiya 3(147),* 26 (1986).
23. Griffith, E.J., *J. Pure and Applied Chem.* 44, 173 (1975).
24. Griffith, E.J., *J. Chem. Eng. Data.* 8, 24 (1963).
25. Tanaka, T., *Ind. Eng. Chem.* 17, 341 (1978).
26. Schilb, T.W., Private communication.
27. Groves, W.O. and Edwards, J.W., *J. Phys. Chem.* 65, 645 (1960).
28. Quimby, O.T., *J. Phys. Chem.* 56, 603 (1954).
29. Maxwell, J.C., *Phil. Mag.* 19, 31 (1860).
30. Kubby, J.A., Griffith, J.E., Becker, R.S., and Vickers, J.S., *Phys. Rev.* 36, 6081 (1987).
31. Lyons J.W., *Phosphorus and its Compounds II,* Ed. Van Wazer, J.R., Interscience Publishers, New York, p. 1683 (1961).
32. Young, R.D., *Encyclopedia Of Chemical Technology,* Ed., Martin Grayson, Vol 2, p. 527, John Wiley and Sons, New York, N.Y. (1978).
33. Savage, Hugh, *Water Science Reviews 2,* p. 67, Ed. Franks, Felix, Cambridge University Press, Cambridge, England (1986). Ratcliffe, C.I. and Irish, D.E. ibid. p. 149.
34. Motooka, I., and Kobayashi, M., *Topics In Phosphorus Chemistry,* Vol 10, Ed., Grayson, M. and Griffith, E.J., John Wiley and Sons, New York p. 170 (1981).
35. Van Wazer, J.R., *J. A. Chem. Soc.* 72, 647 (1950).
36. Griffith, E.J. and Buxton, R.L., *J. Am. Chem. Soc.* 89, 2884 (1967).
37. Brown, W.E., *Environmental Phosphorus Handbook,* Ed., Griffith, E.J. Beeton. A., Spencer, J.M., and Mitchel D.T., John Wiley and Sons, New York, p. 203 (1973).

38. Corbride, D.E.C., *Phosphorus-An Outline Of Its Chemistry, Biochemistry, And Technology*, Elsevier Science Publishing Company, New York p. 556 (1985).
39. Griffith, E.J., United States Patent No. 4,346,028 (August 24, 1982).

CHAPTER

FOUR

Phase Behavior and Cake Formation

A note to the reader: If you are well versed in the phase rule and the determination of phase diagrams, the following chapter may be omitted.

In this section phase diagrams will be discussed as they relate to caking problems. The diagrams are important to caking studies for two different reasons. First, much understanding about the nature of a product may come from a knowledge of its phase diagrams. Second, the phase diagram may be used to find solutions to caking problems. Two types of diagrams will dominate—the two component solid-melt diagrams and the three component aqueous diagrams. These diagrams will lead us to information about the formation of compounds and the formation of solid solutions both in anhydrous melt systems and crystallization from aqueous solutions. These can lead to control of the temperatures of phase transitions or hydrate formation or decomposition. The control of solubility and crystal habit may also be explored while determining the phase diagrams of products of interest.

In Chapter VI, dealing with laboratory procedures and test, ways will be discussed to simply and rather quickly determine phase diagrams of the type usually of interest in caking problems. The products may become so complex that when working with involved formulations as used in some fertilizers or detergents there is no reasonable way to approach the diagrams in an industrial climate but these will also become obvious after some experience in dealing with the diagrams. These complex systems can often be separated and the critical components isolated for study. Also it is not unusual to find diagrams of interest may be found in the literature and piecing several diagrams together may yield much knowledge of the behavior of the system. Examples of how this technique has been successfully employed will be explored.

It should be mentioned that the discussion to follow is not intended to do more than review phase diagrams in light of their relationship to caking problems. The diagrams to be considered are the very simplest and an in-depth knowledge can be acquired from any of several good texts dealing with the subject. The book, *Phase Rule and Heterogeneous Equilibrium*, by Ricci gives more information about phase diagrams than any chemist is likely to ever need in working with most caking problems [40].

PHASE TRANSITIONS

The study of phase transitions can very often be approached by a study of some phase diagrams containing the components undergoing transitions. It is not always necessary that phase diagrams be determined to study phase transitions but when a second component is involved or both the temperature and pressure are considered as variables, even with pure systems, a phase diagram is certainly a logical tool with which to study the properties required to solve a problem involving caking and transitions.

When a solid phase transition occurs, a unique crystalline structure is destroyed and a liquid, gas, or new crystal form is created or the reverse may occur and an amorphous substance, a liquid or a gas, may crystallize. First, attention will be directed toward the Crystal I < --> Crystal II type of transition. In phase chemistry Crystal I is usually the higher temperature form with respect to Crystal II and a Crystal I --> Crystal II transition should imply a cooling transition from a higher to a lower temperature. When a phase transition occurs, most of the physical properties of the crystal are changed while the chemical properties may or may not change. The density of the solid is changed. The solubility at the transition temperature may change drastically. The heat capacity, heat transfer, refractive index, and obviously the x-ray powder pattern will all be different. This suggests that there are many tools that could be used to study a phase transition, and indeed, this is correct. Second, the interest will turn from the systems containing a physical crystal transformation to phase diagrams where liquids, as well as melts, and solid phases are involved in the caking phenomenon.

As usual, nothing is as simple as it might at first seem. In this instance there are several different kinds of transitions that may be observed. In some systems the one crystal form "distills" into the other crystal form directly. In other systems the transition occurs through a hybrid form that is unlike the starting crystal form or the resulting crystal. The hybrid is similar to a series of solid solutions, whose properties change as the concentration of the components are changed.

The types of transitions considered in *chaos* have some parallel in phase chemistry [41]. The transitions in chaos are a change from a structured state to an unstructured state, as when an air mass changes from a stable state to a turbulent or chaotic state, only to establish new structure in the chaotic state. It has been mentioned in Chapter III in the discussion of hydrates and dehydration that the more perfectly a crystal is grown, the more difficult it is to obtain a spontaneous decomposition of the hydrate as the crystal is heated. Most transitions of crystals begin at an apex of the crystal and once the nucleation occurs the transition is usually rapid and may be chaotic to the extent that some crystals have been known to explode when nucleated. This same behavior is probably involved in the development of a latent image on a photographic film. Those crystals of silver halide, which have had but one or two silver ions reduced to silver atoms by a photon of light, are highly susceptible to additional reduction by a developer while those crystals not subjected to "the butterfly wing effect", a photon of light in this case, are difficult to reduce to an image.

With the current interest in chaos the 1943 quotation from R.A. Bagnold is of interest [41]. Bagnold was addressing the flow of dry sand and how it did not cake. It is the objective of most studies dealing with solids to cause them to flow rather than to cake. Millions of tons of solids flowed in the United States during the droughts of the 1930's and turned the Midwestern United States into what became known as the Dust Bowl. "Here instead of finding chaos and disorder, the observer never fails to be amazed at a simplicity of form, an exactitude of repetition and geometric order unknown in nature on a scale larger than that of crystalline structure. In places vast accumulations of sand weighing millions of tons move inexorably, in regular formation, over the surface of the country, growing, retaining their shape, even breeding, in a manner which, by its grotesque imitation of life, is vaguely disturbing to an imaginative mind. Elsewhere the dunes are cut to another pattern—lined up like parallel ranges, peak following peak in regular succession like the teeth of a monstrous saw for scores, even hundreds of miles, without a break and without a change of direction, over a landscape so flat that their form cannot be influenced by any local geographical features." Similar observations are still under investigation. Barndorff-Nielsen published *Sand, Wind and Statistics: Some Recent Investigations* in 1986 [42].

Polymorphic Changes

The published literature dealing with polymorphic changes is extensive and can be complex. If given a full treatment, it can become a quagmire for those not desiring in-depth knowledge. But the area is so fundamental to the caking of solids it cannot be ignored entirely. The subject, as it applies to the caking of solids, will be treated superficially rather than becoming engrossed in a detailed study. The types of transitions to be considered in this section are the type that are responsible for the allotropic forms found as a result of phase transitions of elements. Perhaps the most sought after transition of this type is the conversion of graphite to diamond. The reverse reaction, diamond to carbon, is equally interesting but few investigators have devoted much time and energy to converting diamond into carbon. The author was engaged in a project many years ago in which he converted many grams of diamond into carbon and other products in the study of the reverse reaction. Diamonds burn well in chlorine gas when they are heated in the gas. Fortunately, imperfect and discolored diamonds are relatively inexpensive. Perhaps the current interest in carbon fibers will create an interest in the reverse reaction. It is probably highly significant in obtaining better and better carbon fibers.

Any reaction of a solid, chemical or physical, will result in a phase transition because an existing phase will disappear and a new phase will be formed. Otherwise no reaction has occurred. This statement may seem trite on the surface, but it is the primary consideration in all forms of caking that are not mechanical, plastic, or electrical charge related. It is the type of caking that occurs because either the surface or body of the powdered substances have undergone gross changes.

In the type of transition chosen to be considered, the integrity of the molecules or ions involved in the transition will not usually be violated. A new crystal form will result. Not all transition changes are of this kind. The transformations of sodium metaphosphates are excellent examples. Not only crystal structural changes occur at transitions but a change in molecular structure as well. The polymorphic changes of sulfur are similar. The molecules containing only sulfur atoms change both molecular weight and molecular structure. On the other hand, the many phase transitions of ammonium nitrate involve only the relative placement of the ions or the freedom to rotate in a crystal lattice. In all cases it is ammonium nitrate before and after a solid-solid phase transition. Even the solid-liquid transition yields a liquid that is ammonium nitrate. The system is never more complex than ammonium ions and nitrate ions whether they are contained in a crystal lattice or the fluid of a melt.

Only one type of transition will be chosen as an example relating to caking. Specifically, the first order, discontinuous phase transition will be chosen because it is the simplest, the most common, and the transition type with which the author has had the most first person experience. For the same reasons nitrates and phosphates will often be used as examples. This way the reader is less likely to be misinformed.

Most transitions of the type considered above, but not all, are reversible. However, it is not necessarily true that the down temperature of a transition will occur at the exact temperature of the up temperature transition. If the investigator has not been warned, this can be very confusing when measuring properties for phase diagrams. It is possible to obtain data that scatter widely for a single transition. (To be certain that up temperature transition and down temperature transition are understood if a transition occurs when a crystal is being heated, it will be called an up temperature transition but if the transition occurs while the crystal is cooling, it is a down temperature transition.) There is some inclination for hysteresis to occur when crystals are cycled through a transition. Because of the reversible nature of most transitions to caking, most substances are not stored at a constant temperature. Warehouses change temperatures drastically during the course of a day and a product may suffer several transitions in a twenty-four hour period. Each transition will be inclined to fuse the particles even more tightly.

Not all transitions are spontaneously reversible. A crystal may exist hundreds of degrees above or below the temperature at which it is expected to disappear. Diamond and graphite are a well-known example and the two forms of sodium tripolyphosphate, of detergent and lake fame, are very interesting in this respect. The low-temperature form of sodium tripolyphosphate is easily converted to the high-temperature form when it is heated. But, the high-temperature form may not spontaneously convert to the low-temperature form, unless some amorphous phosphate is contained in the sample. The method of manufacture of a sample can usually be revealed by a study of the behavior of the cooling characteristic of the sample. A sample of sodium tripolyphosphate prepared by tempering glass of a sodium tripolyphosphate composition will typically show a reversible transition as the temperature is lowered. The reversibility may not last for more than one cycle if the minute quantity of glass contained in the high-temperature form is crystallized during the tempering that occurs as the sample is cooled, heated, and cooled once again.

Physically there is nothing discontinuous about a first order discontinuous phase transition. It is a name that differentiates a transition in which the partial derivatives of the free energy are discontinuous. Volume, entropy, and enthalpy are such functions. The second order transitions are the type in which the second derivatives of free energy are discontinuous, whereas the first derivatives are continuous. Curie points and lambda points are of this type. The transitions of the low-temperature forms of liquid helium are of second order type as are the transitions of ferromagnetic crystals to paramagnetic crystals and the transitions of ferroelectric crystals at the Curie point. If it were not for the electric properties of crystals that are discussed in Chapter V, the second order transitions could be ignored in the discussions.

As indicated previously the order of a transition is derived from the mathematical concept, discontinuity. If a function has a value of infinity for some value of the interval of the independent variable, then the function is discontinuous for that value of the independent variable. In the case of the transition, the enthalpy rises rapidly at the transition temperature and the heat absorbed per mole of solid per degree Celsius, C_p approaches infinity over an infinitesimal interval of temperature. The C_p applies to the crystal form that is becoming more and more unstable as it absorbs more and more energy.

It should be noted that all transitions from a lower temperature stable crystal form to a higher temperature stable crystal form absorb heat and are therefore endothermic. Any time that an exotherm is encountered in a heating curve, it means one of two things: 1. an exothermic, chemical reaction has occurred, as perhaps the oxidation of the sample being heated, or 2. the substance being heated is metastable in the temperature range in which it is being heated and transforms to a stable state with the release of the excess energy it contained causing it to be unstable in the first place. The above treatment of enthalpy is meaningful only when the transition is occurring between stable states. If a glass suddenly crystallizes while being heated, one anticipates that an exothermic reaction will be noted because the glass is metastable in the temperature range it is being heated.

A solid-to-solid phase transition, where no new compounds are formed and no liquid phase is formed, may cause severe caking for a variety of reasons. Consider for a moment the interface between the two crystal forms. Assume that a transition of a particle is underway and crystal "A" is transforming to daughter crystals "Bs". There may be only one daughter crystal or there may be many depending on the conditions and the system. Assume for the moment that only one daughter results from the mother crystal and the transition is one half completed. One half of the crystal is crystal "A" and one half is crystal "B". If the energy could be very carefully controlled and no energy was added or subtracted from the crystal, in theory, the two forms could remain in *dynamic* equilibrium indefinitely. At the interface between the two crystal forms, there is a molecular storm that can be easily seen with an ordinary optical microscope equipped with a hot stage and polarized light.

The phase transitions of ammonium nitrate are particularly vivid and are easily seen. A specimen may be prepared by either dipping a microscope slide into molten ammonium nitrate or by heating a few crystals to a melt on a microscope slide, to yield a thin film. When the cooled film of crystals are observed as they pass through a transition, there is then no doubt as to why solid-solid phase transitions can obey the Clausius-Clapeyron equation (Equation 2). The transition is a distillation process. The mother crystal distills across the interface gap to build the daughter crystal. It is seen that this behavior is very important in controlling certain types of caking problems where transitions are involved. If an additive is more soluble in one of the phases than it is in the other phase, it is possible to move transition temperatures many degrees by the addition of small quantities of additives.

If two particles are in contact when one or both of the particles undergo a transition of the type described previously the chances are very good that the particles will cement to each other during the process. This behavior is observed in Figures 7, 8, and 9. Two small particles, each containing many smaller crystal, are slowly heated from room temperature to about 50°C on the hot stage of a microscope while viewing the particles with the dark field of polarized light. As the crystals pass through the transitions, their axes are reoriented with respect to the polarized light. This accounts for the change in brightness of parts of the crystals. It should be mentioned that the transition is occurring more than one hundred degrees below the melting point of the crystals and yet when the transition is completed the two crystals are welded together. The mass transfer that occurs is vividly demonstrated.

In Figures 7, 8, and 9, the photographs were made while the sample was held at a temperature near 50°C. The ammonium nitrate has transformed from Form IV, the room temperature form to Form III. Ammonium nitrate's crystal forms are easily reversible, although there is some hysteresis, and when the sample cooled back to room temperature, the transition was reversed and Form IV returned bonding the particles even tighter.

In addition to the cementing caused by bridge formation of adjacent particles, the mother crystal—daughter crystal behavior can be both interesting and difficult to control in caking problems. Usually in

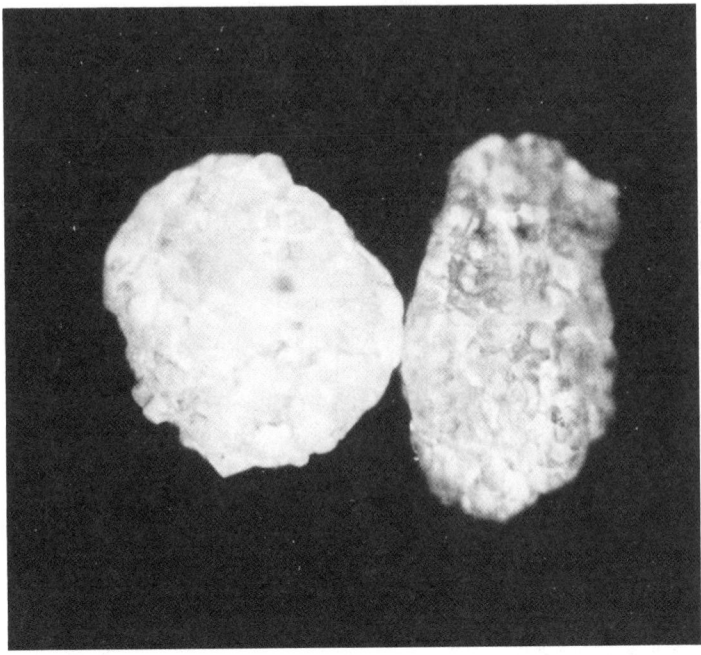

FIGURE 7. *Two small particles of ammonium nitrate viewed with polarized light at room temperature (50×).*

the type of crystals under consideration here not one daughter crystal occurs from a single mother crystal during a transition but hundreds of daughter crystals may form from a single mother crystal. It is also expected that the axes of the daughter crystals will be oriented randomly with respect to the axes of the mother crystal. This alone may lead to a type of caking because this leads to a reduction in the bulk density while the average particle size is also reduced. The decrease in bulk density creates an increase in volume and an increase on the pressure on the contact points within the sample, while increasing their number of particles. If the transition is reversible and the product is experiencing large temperature fluctuations, everything is prepared for the next repeat where the new daughters become mothers and the whole process is repeated. As one might expect, it is possible that the action of the transition is to cause a dust to form that does not form cakes. This is another serious problem that will not be pursued.

A number of excellent works have been published, which deal with the thermodynamics of phase transitions in solids. *Thermal Transformations In Solids* by A.R. Ubbelohde is outstanding [43]. The reader is referred to this work because it is as timely today as it was

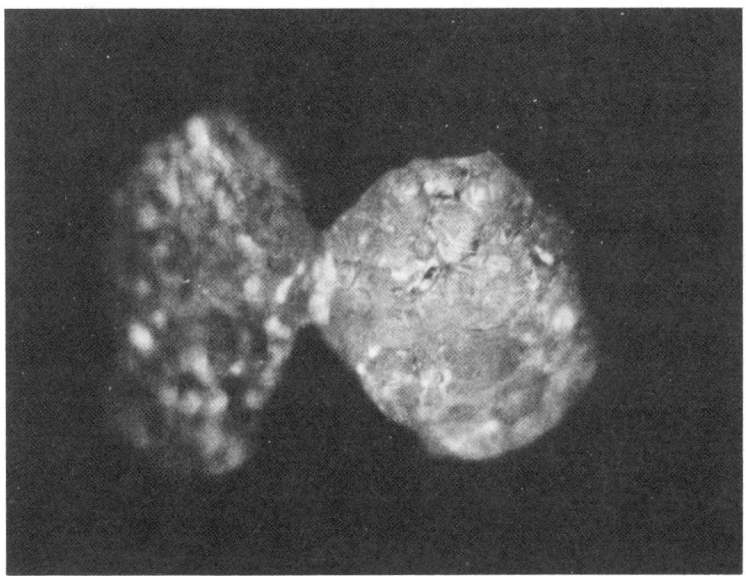

FIGURE 8. *The particles of Figure 7 after being heated through a phase transition neat 50°C on a hot stage of a microscope.*

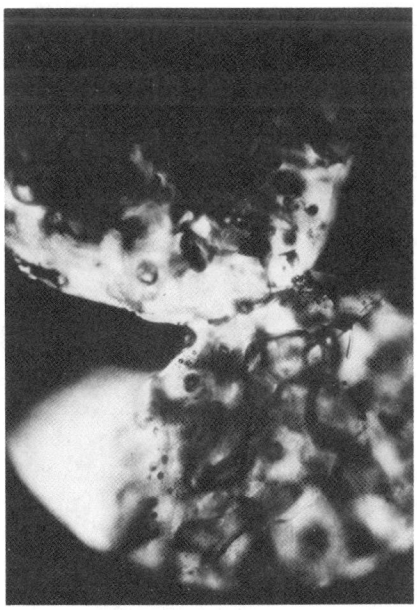

FIGURE 9. *The bridge formed between the particles of Figure 8 after the particles are cooled back to room temperature.*

the day it was written. The mathematical treatments of continuous and discontinuous transitions are informative as to what one should look for in a caking problem. In the discontinuous phase transition it would be expected from theory that in all cases the mother crystal should form many daughter crystal and that the axes of all the daughter crystals should be oriented randomly with respect to the mother crystal. This is seldom the case and often the mother is converted into only one daughter crystal, but the daughter crystal's axis is always directed at an angle to the axis of the mother crystal. This can have a very direct influence on the caking of the crystals, and it is usually advisable to cycle crystals through multiple transitions before passing final judgement on the caking characteristics of a product.

Ubbelohde's article is directed at showing the limitations of the classical Phase Rule as derived by Gibbs. There is small doubt that Ubbelohde is correct but for these purposes the classical Phase Rule is completely adequate and it would probably create more confusion than could be clarified. A very short and interesting discussion of first and second order phase transitions is presented in the *Encyclopedia Of Science and Technology* by McGraw-Hill [44].

Macer and Streeter also discuss the subject of crystal transitions in detail and in some respects their work is more easily read than Ubbelohde's although it is older [45]. It is not necessary to completely master the mathematics and thermodynamics to learn enough of the understanding of the systems to greatly reduce the effort required to obtain a satisfactory answer to most caking problems. The age of these articles illustrates a point that seems to be overlooked by some workers. Outstanding work is ageless and a fact concerning a system, that was a fact during the time of the early Greek philosophers, is a fact today. Do not ignore the older literature when dealing with practical problems.

In *Chemical Identification by Means of Polarized Light Microscopy*, West outlines how the polarizing microscope can be used to the greatest advantage in the work presented here [46]. This subject is discussed in greater detail in the test procedures. Recommendations for both test and equipment are included.

Many techniques to study phase transitions are presented in Chapter VI. Perhaps the two most useful tests employ the hot stage microscope and differential thermal analyses. Between the two methods much may be seen directly about the transition. Very useful equipment was modified with a photomultiplier tube and an x-y recorder to microscope equipped with a hot stage. This resulted in a record of any change in the optical properties of the crystals as a function of temperature. Today the attachment of a television monitor to the microscope is very easily and inexpensively accomplished. The signal from the camera is fed into a television recorder

(VCR) and a permanent record can be obtained along with a voice description if so desired. This approach is currently used.

From the previous discussion it should be reasonably obvious that in the vast majority of cases a caked product has less total energy than the same free flowing powder before it cakes. It is at once noted that the kinetic energy of the powder is potentially greater than the caked material although it may not be so. The consequence may be that some reaction has occurred to cause the powder to cake. If a powdered system were held in an isoenergetic condition, it would never cake but even the interaction of polar crystals could cause some caking on long standing and a quantity of energy would be lost from the powdered system. The reverse is more easily seen. If a system is caked, energy must be put into the system to reconvert it to a powder even if it had been a powder before it caked.

PHASE DIAGRAMS

Understanding caking requires some knowledge of phase science. This is particularly true if one desires to solve problems where caking is involved. In the next few pages several types of phase diagrams critical to many caking studies will be presented. Only two types will be considered. The two component solid-solid diagrams and the three component triangular diagrams in which one of the components is water are considered. Although the systems may be much more complex than this, many of the problems encountered in the types of markets serviced by solid powders may be handled with these two types of diagrams.

A phase diagram is by definition an equilibrium diagram. Many phase diagrams are determined by differential thermal analyses. Obviously one is pushing the concept of equilibrium of a system when differential thermal analysis is used to determine a phase diagram, but if the test samples are large enough and the heating rates are not excessive, thermal analyses can be used to construct reliable phase diagrams. Differential thermal analysis samples that are too small or heating and cooling rates that are excessive may lead to results that are so erratic that a phase diagram may be impossible to construct from the poor data.

DEGREES OF FREEDOM

Degrees of freedom in phase systems are stated in so many different ways it is difficult at times to keep in mind the physical significance of a degree of freedom. It is assumed that only temperature, pressure, and components are variables in the phase rule. The number of independent variables, which must be stated to define a system at equilibrium, is the number of degrees of freedom of the system. The statements are completely general but to have specific meaning they must be applied to a specifically defined system.

A curve in a phase diagram separates two phases. If a one component system as water or sulfur or methyl alcohol is considered, then to remain on a line as conditions are changed either temperature or pressure must be stated. If pressure is known, then the temperature can be read from the diagram. Note that the addition of a component increases the degrees of freedom by one, while the formation of a new phase decreases the degrees of freedom by one. For this, as well as other reasons, the more components used in a formulation, the more difficult the product becomes to control and the more likely the product is to cake.

Two Solid Components Diagrams

Figure 10 is a simple eutectic, solid-solid, two-component diagram. A eutectic mixture, that is the solid that crystallizes at an invariant point between two compounds, has interesting properties in as much as it is a physical mixture of the two compounds but the crystallite size of each compound is usually very small. Each compound maintains its normal crystalline properties, but the solid physical properties of a macroparticle may be considerably different from an admix of the crystals.

As a refresher perhaps it will be helpful to consider Figure 10 in some more detail. It is a two component system. A component is an identifiable, independent chemical species. It may be a compound, portion of a compound, or ions in a salt. To belabor this point, P_2O_5, Na_2O, and $Na_5P_3O_{10}$ may all be considered components but not all simultaneously because there is an equation that equates P_2O_5, Na_2O, and $Na_5P_3O_{10}$, and all three cannot be simultaneously independent.

$$5Na_2O + 3P_2O_5 \longrightarrow 2Na_5P_3O_{10} \qquad [16]$$

Next consider the line *a-e* in Figure 10. This line represents a change in temperature at constant composition and constant pressure and nothing else. This discussion is centered on one single sample of the many that could have been chosen from the phase diagram. It is composition that defines this particular system. The composition may be varied throughout the diagram, but the number of components is fixed for the diagram. Constant composition merely means that the total system has a sum of compositions that add up to 100% when all phases are considered.

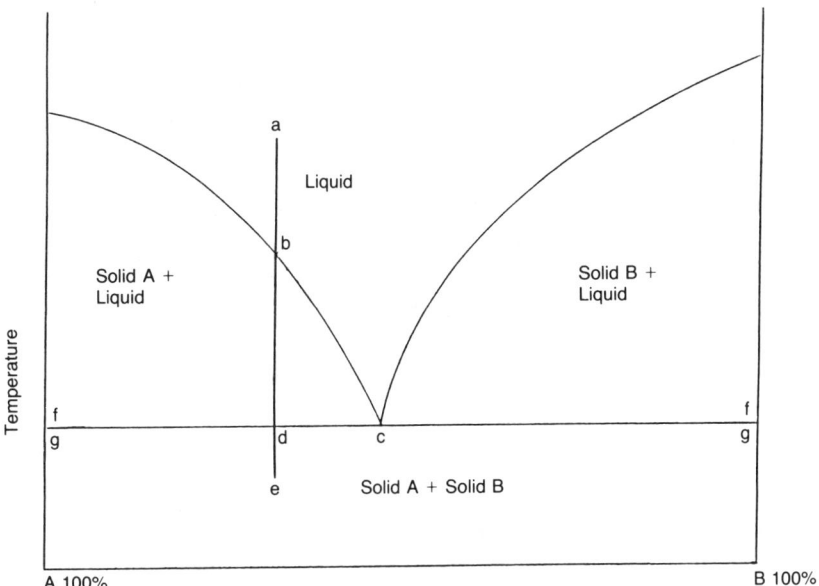

FIGURE 10. *The phase diagram of a simple eutectic system.*

Assume one makes a mixture of A and B with a composition and temperature as represented by "*e*" in Figure 10. If the mixture is heated from *e* to *d* absolute nothing happens as far as phase chemistry is concerned until the temperature reaches point *d*. At the temperature of point *d*, all mixtures of A and B begin to melt and a new phase (liquid) is formed. The formation of a new phase places new restrictions on the system and the system has one less degree of freedom. All independent variables, pressure, temperature, and number of components are now fixed. (Temperature is the only independent variable over which there is a choice because it had been agreed that the system should be composed of two components, and it was agreed that the diagram would be determined at atmospheric pressure.) If the temperature is raised, the B phase is lost by melting into the liquid phase. If the temperature is lowered, the liquid phase is lost. If the system is to remain at point *d*, the temperature must be held at this temperature.

At the temperature of the line represented by *d-c*, three phases exist any place along the line, excluding pure A or pure B (i.e., points *f* and *g*), and with a two component system there are zero degree of freedom as long as the three phases coexist. If one either heats the system until all solid B is melted or cools the system until all liquid *c* is crystallized, the system again has one degree of freedom, but this degree of freedom is temperature. The temperature may be lowered or raised without creating or destroying a phase provided that a phase boundary is not violated. In other words the temperature may be changed within indicated limits without destroying the phase equilibrium that exist.

When the temperature of point d on the Figure 10 diagram is reached, the system is no longer in a two-phase region but has now moved into a three-phase system. This is because A and B begin to melt together to form a liquid, or melt, of composition c. There are now three phases A, B, and melt of composition, c. The system now has zero degrees of freedom. Point c is a unique point in the diagram and no variable must be specified to define the system. But point c is unique only because the solid and the melt have the same composition, and it is the lowest temperature that a melt can exist in equilibrium on the diagram. All points along the line f-g containing point c from pure A, but not including A, to pure B, but not including B, have zero degrees of freedom. From a phase rule point of view, it is the temperature that is unique. There is only one temperature in a simple single eutectic phase diagram at which three phases may coexist in a two-component diagram at constant pressure.

Ideally, as more heat is added to a system at composition d more and more liquid of composition c will be formed and the temperature of the system will remain constant if heat is added slowly enough. If the heat were added in infinitesimal increments, then the system could be held at equilibrium and the temperature would remain constant. The temperature will remain constant at the temperature of d until all of the component B has melted and only solid A and a liquid of composition c remains. Now the temperature begins to rise as more heat is added because the system now has one degree of freedom and only solid A is melting. If the temperature is known, then the composition is known from the diagram *at equilibrium*. The melting of A into the melt containing B is unlike the melting of A into a melt of pure A if it assumed that both A and B melt congruently. In the first instance the temperature continues to rise because the composition of the melt is continuously changing. In the second case the first melt formed is exactly the same as the last melt formed provided the compound does not decompose in an incongruent melting.

As more heat is added (i.e., the temperature increases) to the two phase, solid A + liquid, system of overall composition, e, the composition of the melt follows the curve from c to b. Note that the temperature and composition of the melt approach b there is less and less crystalline solid left in the melt and if all has performed perfectly at point b there is no solid left in the melt. In the determination of the phase diagram, the exact location of point b is very important. If heating curves are being utilized with a differential thermal analysis, it is easily seen why the melting of the last trace of crystal may be difficult to determine. Conversely, when cooling curves are used, if the system super-cools at all and they usually do, the temperature of formation of the first crystal may again be difficult to obtain. Methods have been developed to overcome these problems by approaches other than differential thermal analysis, but their utilization in industrial laboratories is usually too time consuming to consider in detail. Point *a* on Figure 10 is of course a one-phase melt with two degrees of freedom.

Before leaving Figure 10 a discussion of the types of solids that are likely to form a system having the simple eutectic point in the diagram is in order. As liquid crystallizes at the eutectic point, both crystals of A and crystals of B co-crystallize. This means that the solid that is formed is very likely to be composed of a mixture of very fine crystals. If the solid is put into water and a three component diagram is determined, the eutectic behavior may or may not continue to dominate. The two crystal forms that make up the eutectic then may or may not have similar solution properties. If one crystal is much more soluble than the co-crystal, then both crystals may show an exaggerated caking behavior. A last point here is to remember that the x-ray pattern of the solid crystallized at the eutectic point will be a mixture of the patterns of each pure crystal form. The crystals are very small and intermixed.

In Figure 11 a similar type of diagram can be important to caking problems because the two starting compounds, A and B, crystallize to form a new compound C, a double salt, with properties unlike either of the parent compounds. This is easily seen by a simple x-ray diffraction study. The new compound will have a new pattern unlike either parent while in the case of the eutectic in Figure 10 the x-ray pattern would be a mixture of both of the starting compounds.

Whether or not a new compound may be useful in solving caking problems depends upon a large number of considerations. In some phase compounds the anions retain their identity while in other systems an entirely new anion can be formed. The silicate and phosphate systems have many examples of both types. The salt $2NaH_2PO_4 \cdot Na_2HPO_4 \cdot 2H_2O$ is a double salt from an aqueous system where the anion retains its identity while $Na_5P_3O_{10}$ is an example of the compound that is formed by heating $2NaH_2PO_4 \cdot Na_2HPO_4 \cdot 2H_2O$ and Na_2HPO_4 above their decomposition temperatures.

In the detergent industry the most common cause of caking is a high water solubility of one of the components in a formulation. If one can form a double salt between the required component and a second similar component where the double salt has less solubility

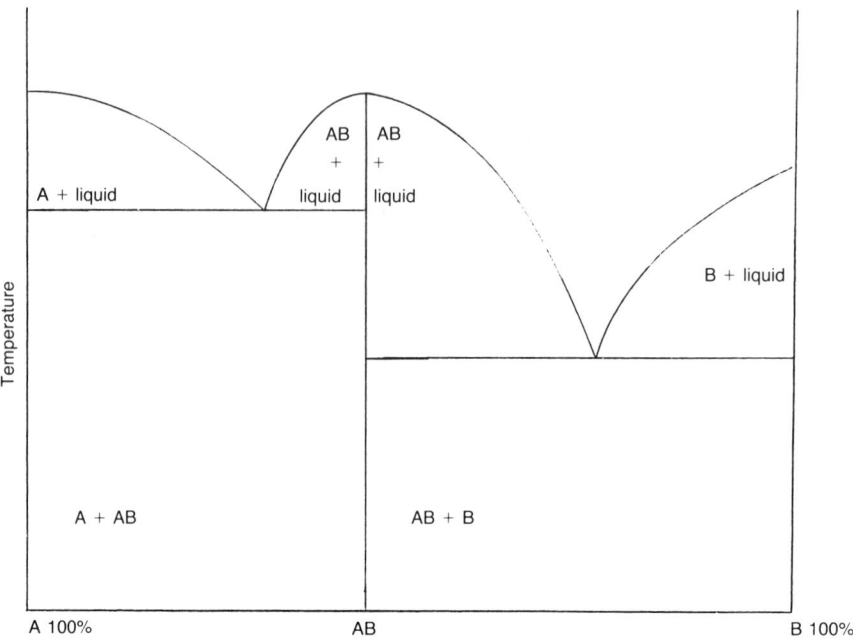

FIGURE 11. *The phase diagram of simple double salt formation.*

than the required component, it is sometimes possible to eliminate a caking problem. There are a number of restraints, which must be recognized, if the double salt approach is to be practical. It would be desirable that the double salt should contain several molecules of the desired component for each molecule of the additive. Although some double salts of this type exist they are relatively rare. The second requirement is that the additive contributes functionally to the system, or as a minimum, does not subtract from the functionality. These requirements are sometimes found with mixtures of optical isomers. All manner of unexpected behavior may be found in these systems. Menthol is an outstanding example. It has several isomers and a double melting point. On cooling liquid menthol it will freeze to a solid and when the temperature is lowered it will again melt and refreeze.

Finally, it should be noted that Figure 11 may be considered as two diagrams of the Figure 10 type that are placed back to back. In fact one could daisy-chain one diagram to another and so forth. The only requirement is that the end component is common to both connected diagrams. If phase diagrams were daisy-chained the connecting compounds may or may not be double salts of the type shown in Figure 11. Yet another way of considering this is to realize that it is possible to have "n" diagrams, radiating in "n" directions from a single axis. Again the only limitation, for a two-component system, is that all connecting diagrams have a single component in common at the axis of intersection.

Perhaps the most powerful of all tools for modifying the properties of one substance with small quantities of a second substance is the solid solution. Sometimes great changes in hygroscopicity, solubility, hardness, and so forth may be obtained by minor additions of a second substance to the primary compound. It is always worth mentioning that a solid solution is not a solution at all, but a crystalline solid composed of two or more components in which the components behave as though they were uniformly dissolved in each other. The properties change in a continuous manner as the composition is changed.

There are many types of solid solutions phase diagrams but only a few will be mentioned. Figure 12 represents a continuous system of solid solutions with a maximum melting point, or region. Similar diagrams exist where there is a minimum in the curve, while others flow smoothly from one axis to the other.

Consider the point "a" on Figure 12. It should be a clear melt

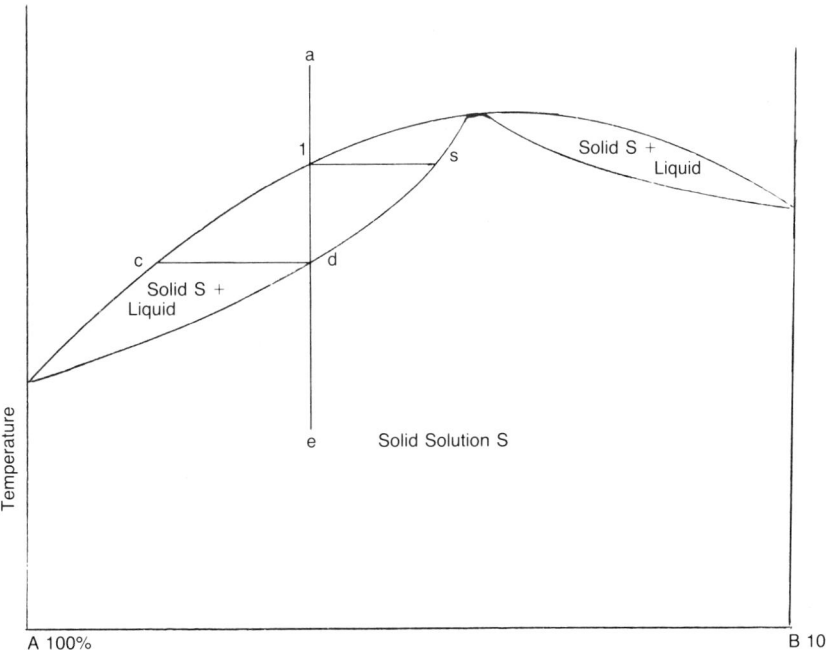

FIGURE 12. *A phase diagram of a system of solid solutions with a maximum melting point.*

with two degrees of freedom. Again, pressure is understood to be held constant in this diagram. Temperature and composition may both be changed and no new phases will be formed, provided the top line is not intersected. If the temperature is lowered from a to b crystals of composition, s will begin to form but only a very small quantity will form of this composition. This is because the crystals are richer in B than they are in A. The liquid phase is loosing B in greater amounts and the liquid melt composition moves toward c. The liquid phase is moving toward c continuously as the temperature is lowered the solid crystals that are forming are getting richer and richer in A. It should be obvious that it is difficult to purify the components of solid solutions by crystallization. It may be done but usually requires several recrystallizations. At the maximum in the curve, point g, it is possible to crystallize a solid solution that has the same composition as the melt.

Figure 13 is similar to Figure 12 in many ways but is very different in others. Two solid solutions are formed in the systems. The two solid solutions may be very different in properties, and they are not soluble in each other. Consider the point with temperature and composition represented by "a". As this system is cooled crystals built on

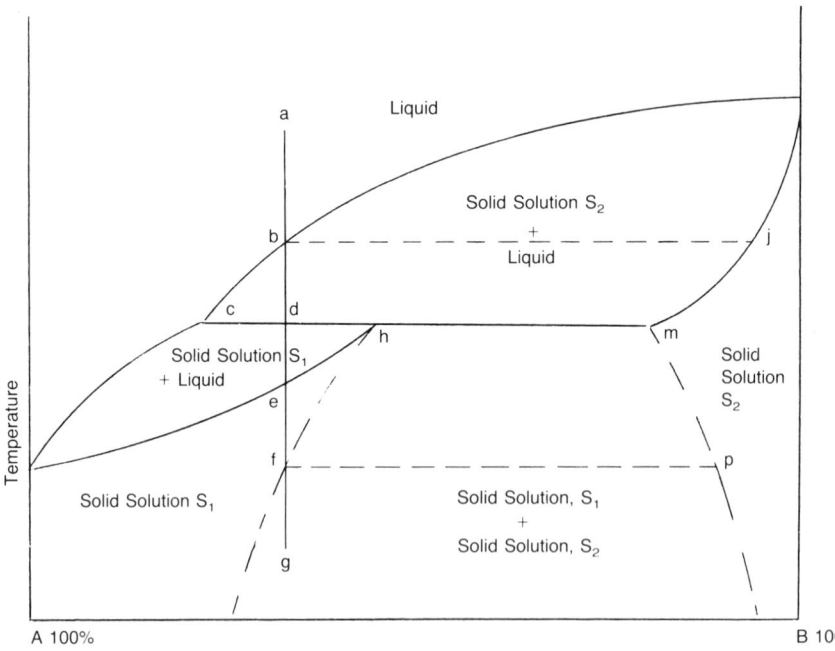

FIGURE 13. *The phase diagram of a system with two solid solutions with a miscibility gap.*

the S_2 structure will begin to crystallize at b and will have a composition, j, much unlike the melt from which they crystallized. At point b there are two phases, two components, and therefore, one degree of freedom. As heat is removed from the system, the liquid phase moves toward c while the solid moves toward m. At point d liquid of composition c is reacting with solid of composition m to form a new solid of composition h. Since there are now three phases and two components, there are zero degrees of freedom and the temperature will remain constant until all of the solid m has been reacted to make a system of solid solution S_1 with a composition h. As more heat is removed, the liquid phase continues to react with the solid and the solid composition moves from h toward e. At point e the total systems becomes solid. One should think that is enough changes for one sample but the changing is not quite over. Between points e and f only the solid S_1 exist but at the temperature of f the solid solution S_1 begins to fall apart to a mixture and the solid solution lost on the line c-m is back again. **BUT** it must be recognized that although the second solid solution formed at point f is based upon the structure of S_2 its composition starts at point p and as the temperature is lowered more S_1 reacts to form more solid solution, S_2. The only thing in the system that ever remains constant is the total composition.

With the previous background and the seemingly never ending interest in ammonium nitrate, as judged from the number of articles in the current literature, the phase diagram for the NH_4NO_3-$MgNO_3$ system will be discussed in some detail [47]. It is a classic example of how caking of ammonium nitrate may be completely avoided under ambient conditions provided the system is not allowed to absorb water from the atmosphere.

It was at first recognized that if *soft burned* magnesium oxide was dusted on the surface of either powdered or prilled ammonium nitrate, even that which was made from hydrous melts, that some ammonia was liberated but that the product became dusty and would remain free flowing indefinitely provided it was given the normal humidity protection required by ammonium nitrate, because of its very high solubility. This suggested that magnesium oxide added to an anhydrous melt should have some interesting properties. It cannot be overly emphasized that there are enormous differences between hard burned and soft burned MgO or hard and soft burned CaO. If hard burned MgO is dusted over the surface of ammonium nitrate prill, little or nothing will happen for a long time.

It was soon recognized that if as little as 0.5% MgO was added to ammonium nitrate melt when prills were formed the transition, which normally occurred near 32°C, could be eliminated in the temperature range below 50°C. In this case hard or soft burned MgO can be used in the melt but the reaction is faster with soft burned magnesium oxide. Experimental prills were prepared from anhydrous melts. The finished prills had the appearance of ceramic spheres and they were very hard and smooth. The prill would not reside on sloped belts usually used to transport prill from one point in the manufacturing plant to another but would roll back to the base of the coolers. Front end loaders that normally were used to lift and transport prills from one spot to another would not pull into a pile of prills but would sit on the prills with their wheels spinning on the ball bearing-like spheres.

The phase diagram for the system was determined as well as the influence of water on the temperature of the phase transition between Form IV and Form III ammonium nitrate. See Figure 14. It is obvious that the temperature of the transition of ammonium nitrate is a function of the quantity of water contained in the ammonium nitrate. All of the additives, which have been added to ammonium nitrate that raise the IV <=> III transition temperature, are merely acting as

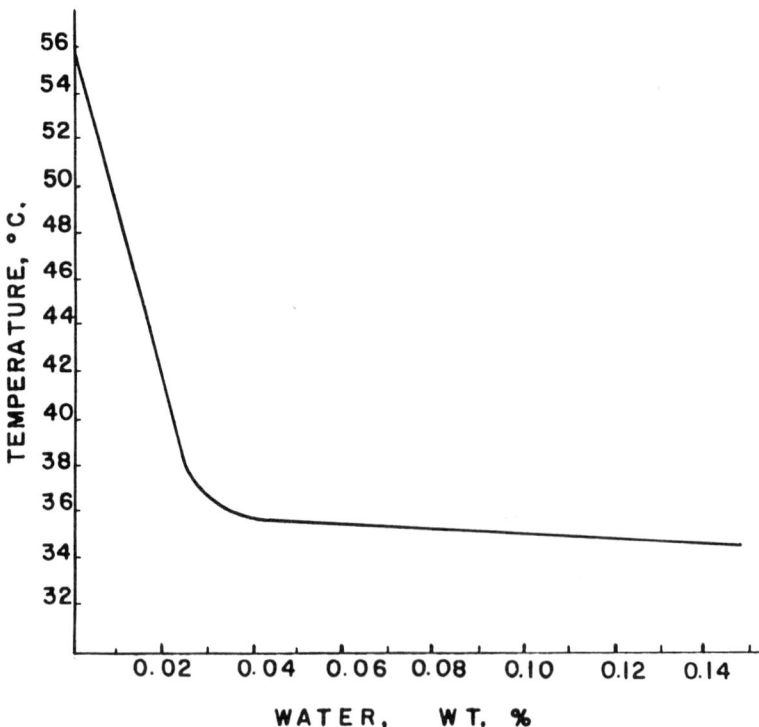

FIGURE 14. *The temperature of a solid-solid phase transition of ammonium nitrate as a function of water content.*

dehydrating agents. These compounds include polyphosphates, boric oxide, calcium oxide, zinc oxide, magnesium oxide, and so forth. The superiority of the additives depend upon the types of compounds formed and the nature of the eutectics in influencing prill properties. The behavior is in some ways similar to the change of alloy properties in metallurgy.

Magnesium nitrate is a magnificent dehydrating agent. Pure anhydrous magnesium nitrate has never been prepared because it decomposes before the last of the water is removed from its hydrate crystals. It may be seen in the phase diagram that the double salt of ammonium nitrate and anhydrous magnesium nitrate may be prepared very rich in anhydrous magnesium nitrate. It should also be known that the systems very rich in magnesium nitrate may start to self-heat at temperatures approaching melting and the hazard of explosion is very likely at high temperatures. The very rich magnesium mixtures should be superior to pure ammonium nitrate for rocket and blasting work however. This is because the percent oxidizer is greater in $Mg(NO_3)_2$ than it is in NH_4NO_3. If the amorphous glass is formed, then the energy of crystallization is also contained in the glass increasing the energy of the system even more.

Consider the phase diagram in Figure 15. First note the rise temperature of the IV <=> III transition as more and more magnesium nitrate is formed in the system. This rise in temperature is probably an artifact because as more magnesium nitrate is formed the system becomes more anhydrous. This could easily account for the rise in temperature. More important is the behavior of the melt as it freezes.

Consider a sample of ammonium nitrate that contains about 0.5% magnesium oxide or about 1.8% magnesium nitrate formed in situ from added magnesium oxide. Now heat the system to a temperature near 180°C. Next let us consider what happens to a droplet as it fall in a prilling tower. Energy is lost very rapidly from the droplet and the temperature begins to fall. The first phase change to happen to the droplet is the crystallization of part of the melt to pure Form I ammonium nitrate. Keep in mind that the initial crystals are pure Form I ammonium nitrate because crystallization may be a highly superior purification technique. As the temperature continues to fall more and more Form I crystallizes until the temperature reaches about 125°C at which time all of the Form I is converted to Form II. More pure liquid ammonium nitrate continues to crystallize while the melt becomes richer in magnesium nitrate-ammonium nitrate double salt until the temperature reaches about 115°C. At this temperature the double salt, $(NH_4)_3Mg(NO_3)_5$, and ammonium nitrate crystallize as the eutectic. The eutectic liquid was pushed to the outside of the

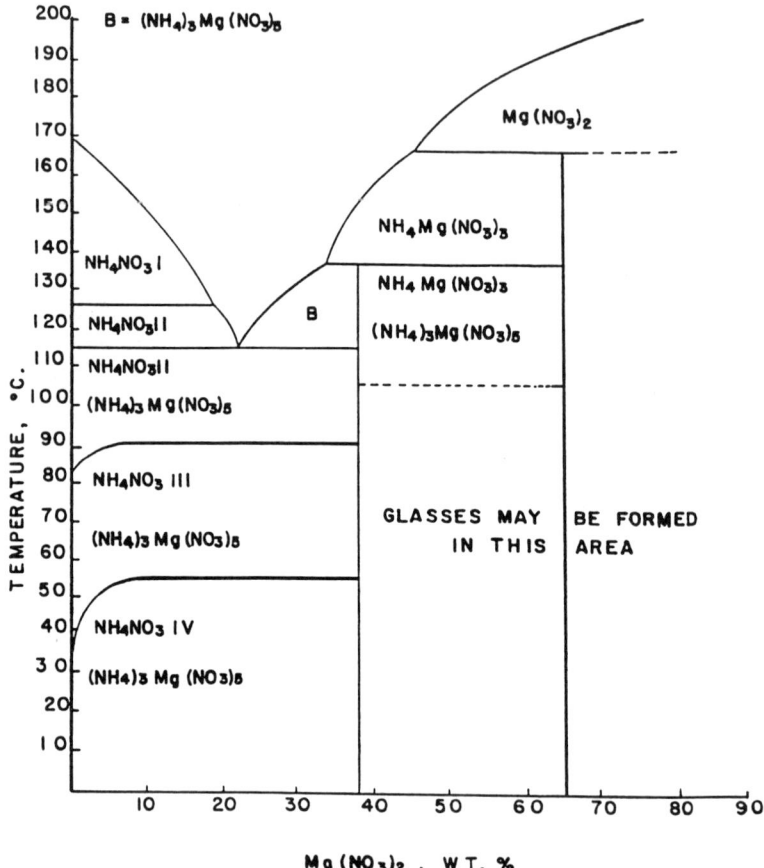

FIGURE 15. *The phase diagram for the ammonium nitrate-magnesium nitrate system.*

prill during the entire crystallization process. It crystallizes last causing the surface to be rich in magnesium nitrate and filling the surface with very fine crystals. This leads to a highly superior prill.

Three-Component Three-Dimensional Diagrams

The three-component three-dimensional phase diagram is one in which temperature is a variable and is plotted on the vertical axis while the base of the diagram is a triangular composition diagram. The liquidus curves are plotted as the surface of the diagram. It is very interesting and useful that these diagrams may very often be constructed from published two-component diagrams, lending much information about a total system of interest when no complete three-component diagram exists. In this way the wall of the diagrams are constructed and often even diagrams that build the internal structure of the diagram have been published.

The following diagram will be used as an example because it served very well in a program that was taken through pilot plant studies. In this case the genealogy of the diagram is known. The three components to be used are Na_2O, CaO and P_2O_5. Four published diagrams two-dimensional diagrams were used to construct the three-dimensional diagram. The Na_2O-P_2O_5 diagram was published by Patridge, Hicks, and Smith as well as by Morey and Ingerson [48] [49]. The two diagrams agree very well in all details. The CaO-P_2O_5 diagram was published by Hill, Faust, and Reynolds [50]. The back wall of the diagram was the $Na_4P_2O_7$-$Ca_2P_2O_7$ published by Berak and Zmanmievowska [51]. A partition wall of the compounds $[NaPO_3]_n$ and $[Ca(PO_3)_2]_n$ was first published by Morey and corrected by Gremier, Martin, and Durif [52] [53]. The composite diagram, Figure 16, was published [54]. With this much information as to how the system behaved, it was not difficult to fill in areas where diagrams had not been published and often a guess proved to be adequate for most preparative work. For example it was desired to grow a calcium metaphosphate composition in a region of the phase diagram that would allow a soluble sodium rich matrix to be extracted. Because it was known that the ultraphosphates degraded easily and were difficult to crystallize, it was logical to explore a region of the diagram where the calcium polyphosphate $[Ca(PO_3)_2]_n$ dominated the diagram as the crystalline phase. This region was identified and fibrous crystals of the desired composition resulted.

The approximate diagram is shown in Figure 16. The diagram is not drawn to scale and the section rich in P_2O_5 has never been determined because much of the composition has never been crystallized. The ultraphosphate $Na_2P_4O_{11}$ has been claimed as a crystallize salt but has then been withdrawn. There is no doubt that the calcium analogue has been crystallized and an amorphous composition of the sodium salt is surely possible.

Three Component Aqueous Diagrams

As aqueous phase diagrams are considered, those that form solid solutions may be potentially very important in solving caking problems caused by highly soluble and hygroscopic salts. At times no other practical course is evident. The next diagrams are of the three component type where water is one of the components. Rather than measuring melting temperature, solubility is measured while both

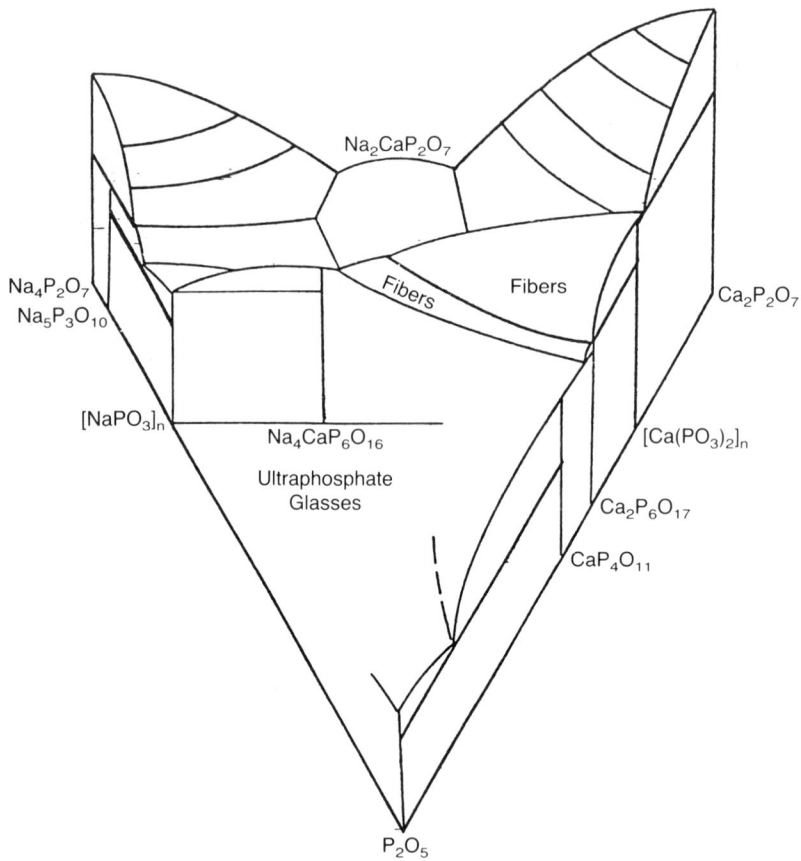

FIGURE 16. *A representative phase diagram based upon previously published two component diagrams.*

temperature and pressure are held constant. All examples are plotted on triangular graph paper. Rectangular graph paper is equally useful but in that case two of the components are plotted and the third component is obtained by the difference from 100%. Either weight percent or mole percent of the compounds may be used but in most diagrams the authors use weight percentages.

The simple eutectic type of aqueous diagram is seen in Figure 17. The invariant point in a simple eutectic type diagram is not usually referred to as a eutectic but as a cryohydrate point. The system is composed of two salts that do not react to form either double salts or solid solutions. At the temperature chosen to determine the diagram, compound A does not form a hydrate, and therefore, the tie lines for the two-phase area reach all of the way to the lower left-hand corner of the diagram. Compound B, on the other hand, forms a hydrate and

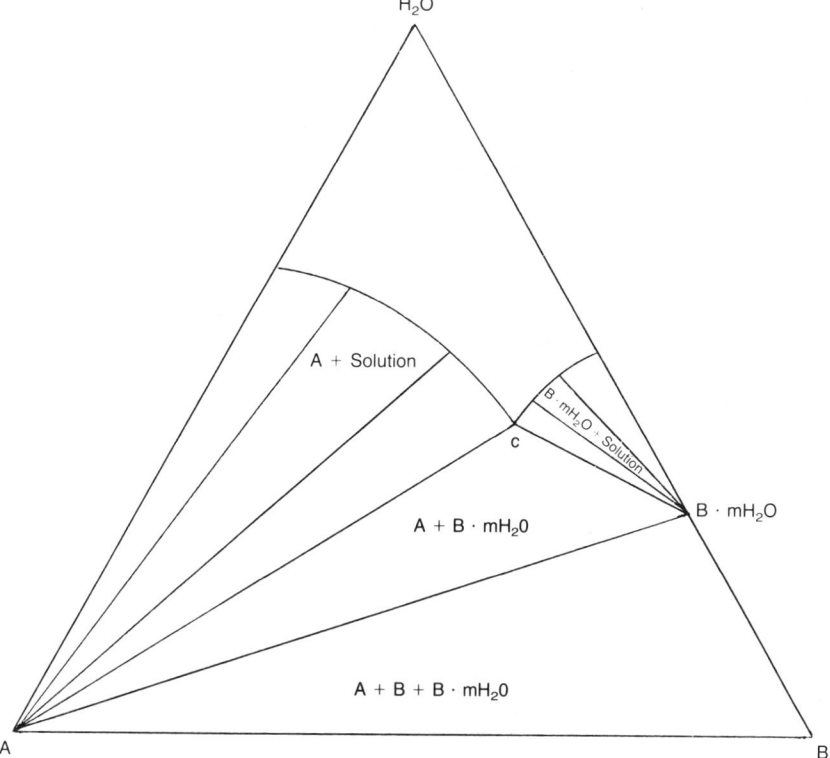

FIGURE 17. *A three component aqueous phase diagram with a single invariant point.*

the tie lines radiate from the composition of the m-hydrate. In this case the choice is to plot the data as weight-percent and the point on the diagram representing the hydrate is merely the weight loss on ignition of the solid hydrate. It is assumed that the anhydrous salt does not decompose.

Let m-2 on the line on Figure 17 that extends from the anhydrous compound to the dihydrate, $B \cdot 2H_2O$ separates dry salts from mixtures of salts and solutions. All compositions below the line are merely dry mixtures of hydrated crystals of B and dry crystals of A. All compositions above the line bounded by the lines from A to the invariant point c to the dihydrate $B \cdot 2H_2O$ are wet crystals in equilibrium with a solution of composition c. This is to say that any mixture of A and $B \cdot 2H_2O$ desired may have the composition c, provided too much water is not poured on the crystals to dissolve them all. The solution would then have a composition somewhere in the one phase area above all of the two phase lines. The solution's composition would depend upon the chosen mixture of salts before water was poured onto them.

All that is now left is the two fan-shaped areas on either side of the diagram. These are two-phase areas. On the left only crystals of pure A and a solution containing dissolved A and B but no crystals of B. On the right the reverse is true. There are crystals of pure B·2H₂O and a solution of dissolved A and B but no crystals of A. The crystals that grow from a solution containing A and B of the right composition to pass through *c* are usually rather fine but they do not have to be as in the case of the eutectics of melt chemistry. In either case large crystals of either compound may be added and equilibrium may be established provided a phase boundary has not been violated.

Figure 18 is a diagram representing double salt formation. Assume that one is interested in using compound A in a dry blend. It is soluble to the extent that 65g will dissolve in 35g of water at the

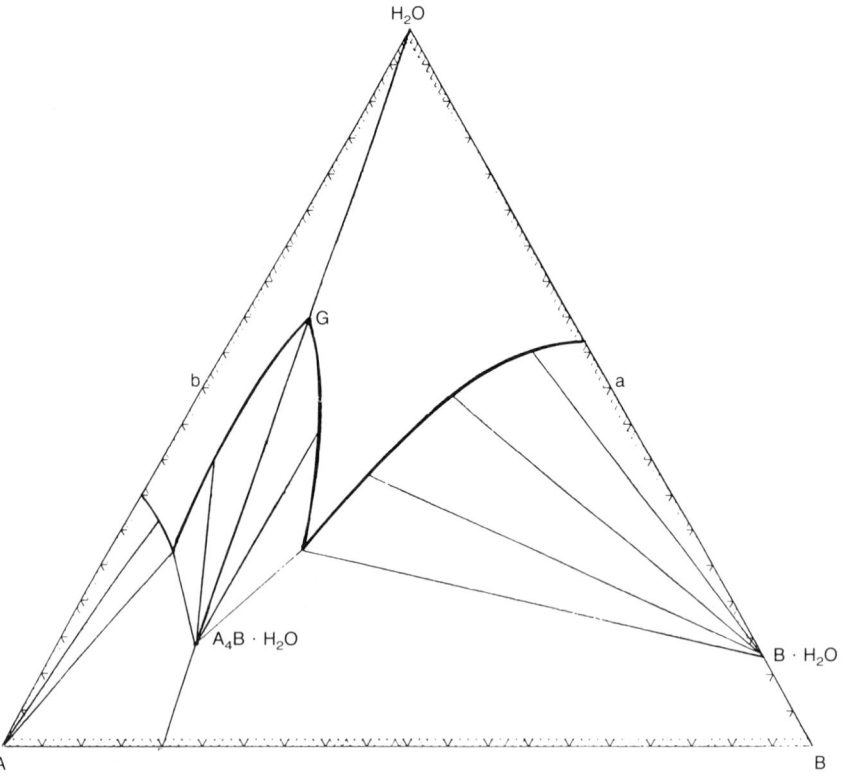

FIGURE 18. *A three component aqueous phase diagram with compound formation.*

temperature of the diagram. The system almost sucks up water. If a compound as B may be found that is acceptable in the product, the solubility can be reduced to 40% by adding twenty mole percent to A. This is still far from ideal but it is much better than before. In addition, the double salt has some water-holding capacity because it forms a hydrate. This will result in two other differences. *All salts in their highest hydrated form dissolve endothermically. All hydrates form exothermically.* When a water molecule in the atmosphere is absorbed by component A, the surface cools if the heat of solution is larger than the heat of vaporization of water, as calculated on a per water molecule basis. Assume that the double salt, $A_4B \cdot H_2O$, also exists. When a water molecule is absorbed from the atmosphere, the surface definitely heats because both the heat of hydration and the heat of condensation are both exothermic. Everything else being equal, and it seldom is, salts that form hydrates have some built in protection. The very first molecule of water to stick on the surface of a salt that does not form hydrates creates a saturated solution on the surface of the salt. Therefore, surfactants are sometimes used on salts that contain voids. Ammonium nitrate prills that are grown in towers from a saturated solutions contain void spaces as do most agglomerated granules. The surfactant prevents solutions from accumulating on the surface of the salts because the lowered surface tension allows the capillary action to draw the water back into the interior of the particles where it is considered to be less harmful. Although it will work for some systems it is not recommended other than as a last resort.

In Figure 19 component A is an experimental detergent builder. Component B is sodium carbonate. It was known that the builder was less hygroscopic when mixed with sodium carbonate. As may be clearly seen in Figure 19, the builder forms solid solutions with sodium carbonate. The solid solution is less soluble than the builder. (It should be mentioned that as the composition changes along a straight line from the apex of the diagram toward the base, water is extracted from the system, usually by evaporation. Conversely, if water is added to the system, the composition will move along a straight line directed toward the apex.) Consider the Figure 19 at point F. It is a clear solution. If water is evaporated from the solution, the total composition will follow the line F-D. At B crystalline salt will begin to form with the composition G. The tie lines connect the solids with the solutions with which they are in equilibrium. As more water is evaporated, the solid solution composition will move from G toward D while the solution will move from B toward E. The solid of composition D is in equilibrium with a solution of composition E. This means that the solid solution D is less soluble than the compound A. The solid solution of composition D can be exposed to a higher relative

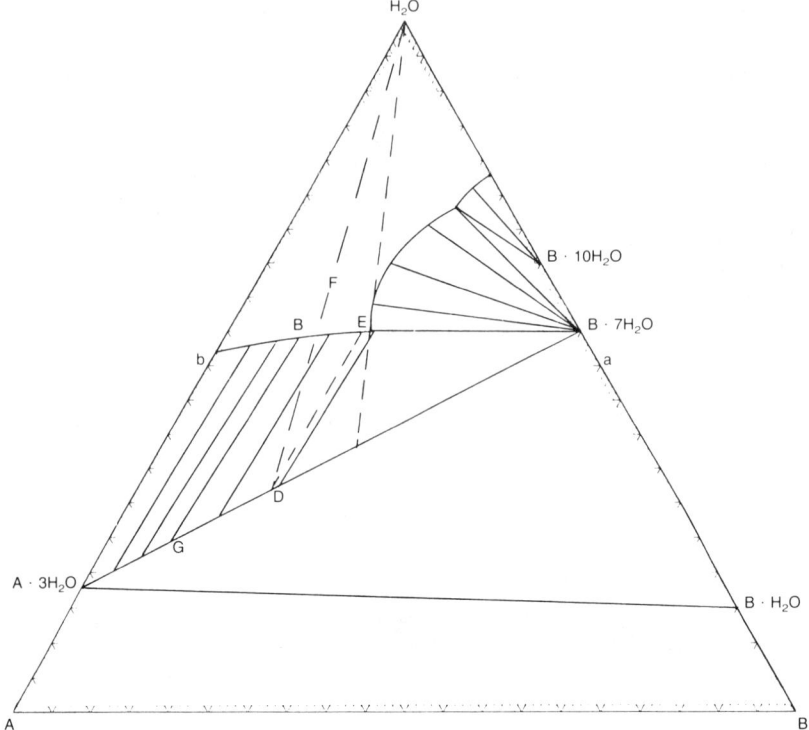

FIGURE 19. *A three component aqueous phase diagram exhibiting both solid solution and hydrate formation.*

humidity than the compound A·3H$_2$O before it begins to absorb water from the atmosphere. If the dry solid solution D is subjected to water vapor, tensions great enough to cause it to absorb water from the atmosphere the reverse of the evaporation will occur. The solid solution is noticeably less hygroscopic than the pure salt both in laboratory tests and pilot plant observations.

If the sodium carbonate had been even less soluble and the solid solution had extended all of the way across the diagram, it might be expected that the differences in hygroscopicity could be dramatic. The question arises as to whether or not some other substance might be substituted for the sodium carbonate? The substance should have much less solubility than the carbonate and it should fit into the crystal lattice of the builder. The builder is a polycarboxylate and it was reasonable to assume that sodium carbonate might fit into the crystal lattice. To find a second salt that would behave the same and fit into the lattice, it was natural to investigate the properties of other carboxylates. Sodium oxalate was an ideal candidate. It had a very low solubility and had an excellent chance of forming solid solutions with the builder.

In Figure 20 when it was at first discovered that the builder formed solid solutions with a salt that was much less soluble than the builder, it was hoped that the solid solution would be much less hygroscopic than the builder but it was not. Figure 20 is the phase diagram for the same detergent builder and sodium oxalate. Sodium oxalate forms a complete series of solid solutions with the builder. It does have dramatic influence on the solubility of the detergent builder. Return to Figure 19 for a moment and compare it with Figure 20. Note the difference in the slope of the tie lines. In Figure 19 the reverse of the behavior occurs. Consider a solution B in Figure 20. When it is evaporated, the first crystals to form have the composition F. As the solution is evaporated more, the solid phase changes along the line from F toward D and the solution phase moves from C toward E. When the last free water is evaporating to give a dry sample, the sample will have the composition D and the last solution will have

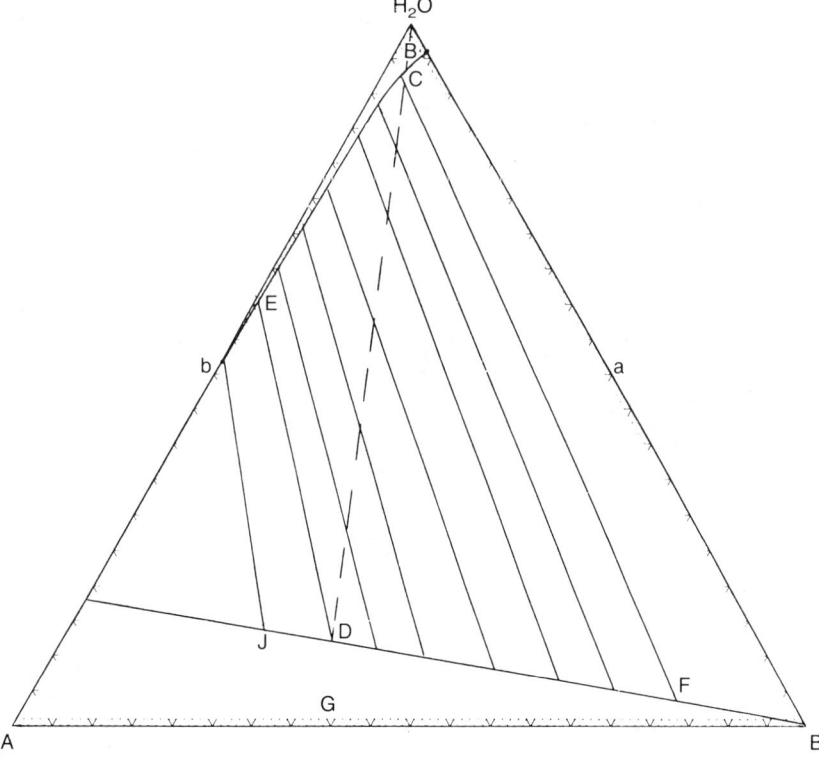

FIGURE 20. *A three component aqueous phase diagram exhibiting a continuous series of solid solutions.*

the composition E. The concentration of the solution at composition E is almost as great as the builder itself.

Assume that a detergent builder containing sodium oxalate has been prepared by driving water from a composition B and the dry solid solution D was formed as a product. If the solid solution is exposed to atmospheric water, it will absorb water at almost the same rate as builder itself. The reason is very simple. The first solution to form on the surface of the crystals of solid solution is very concentrated, having a composition near E. It would, therefore, have a very low vapor pressure and would continue to absorb water until a solution was formed that had a vapor pressure of the vapor tension of water in the atmosphere. The system will follow exactly the reverse path as when the water was evaporated and an ultimately very dilute solution will form. To lower the hygroscopicity of a salt of interest, it is not sufficient that a solid solution may form between the salt and a second salt of lower solubility. If water absorption is to be eliminated the solid solution must behave as Figure 19 and not as Figure 20. The tie lines in the phase diagram of Figure 20 slope strongly in the direction of the more soluble salt. It would be possible to lower the hygroscopicity of the builder by adding enough sodium oxalate but the system should be richer in sodium oxalate than in builder. For the concept to be viable, it is necessary that a small quantity of the less soluble salt significantly lower the concentration formed on the surface of the crystals when they are exposed to humid atmospheres.

Figure 21 is a phase diagram similar in many respect to the diagram presented in Figure 20, previously shown. It is know that oxamide has a dramatic influence on the caking properties of urea. See Chapter VIII. The phase diagram was determined to test whether or not the system contained solid solutions. It was determined by the method outlined in Chapter VI. In this case it is necessary to determine only the percentage water in the solutions. Note that the tie lines all converge to a single point and oxamide is so sparingly soluble as to coat the more soluble phase, urea. This accounts for the behavior of the system as it is discussed in Chapter VIII.

There are several points concerning the use of solid solutions to solve caking problems that must be clearly understood if they are to be utilized correctly. This approach is probably unique to this book. It is a very powerful tool.

Normally the base line on a three-component aqueous phase diagram represents mixtures of dry solid components. If either component forms hydrates, there will also be a line extending from the hydrate composition to the composition of the other component or double salts if they form. When solid solutions are formed, these base conditions cannot be considered as mere mixtures but must be equi-

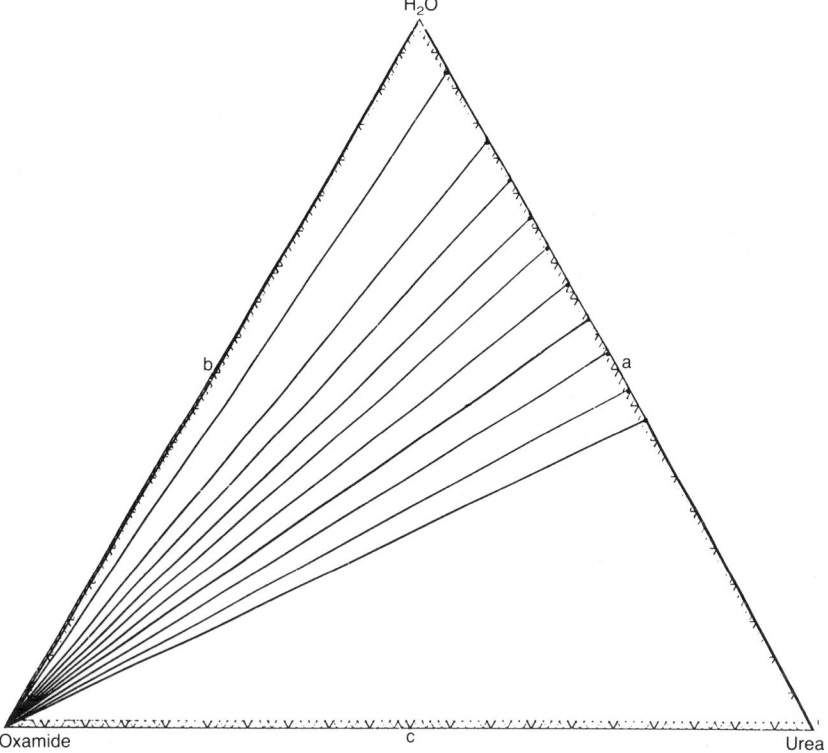

FIGURE 21. *The aqueous urea-oxamide phase diagram at 30°C.*

librium compositions. This is because there are an infinite number of unique crystalline solids along this base line, each capable of expressing its own tie line. Therefore, if the system is to behave as desired, it is absolutely necessary that the solids be prepared in a manner to allow the solids to dissolve in each other. This may be as simple as preparing a slurry and stirring it until equilibrium is established or in some cases it may be necessary to grow the solid solutions from clear aqueous solutions.

Considerable time has been spent discussing the behavior of these systems because they are among the more important concepts presented in this book when practical caking problems are to be solved in systems that are not too complex. Both the fertilizer and detergent areas can apply the concepts well with the solid detergents of the future being prime candidates. Many of the detergent builder candidates of the future will be very soluble when compared to the sodium tripolyphosphate used in the past. This is certain to create caking problems, as is already seen in some brands of detergent that have been free of caking for many years but are now caking when left open over a washing machine.

To use the solid solution formation to lower the hygroscopicity, and consequently, the caking of some solids there are a few rules of thumb that shall be repeated.

1. Check the solubility of the compound in question. If it has never been determined, then determine the solubility of the compound in water.
2. Are other compounds known that function similarly to the compound in question?
3. Are any of the other compounds of similar size and shape to the compound in question?
4. Are any of the other compounds much less soluble in water than the compound of interest?
5. Are the other compounds of similar cost to the compound in question?

If a compound can be found that meets all five of the last requirements, it is probably a good candidate for a trial study. Prepare a trial phase diagram as instructed in Chapter VI.

SUMMARY

Phase science plays an important part in the behavior of solids in all systems. Both chemical and physical changes can lead to caking and a phase change is usually involved in both types whether it is the recrystallization of a component in a product or a chemical reaction which may be a decomposition of all or part of a system.

Two general types of phase diagrams have been considered. The two-component solid-melt diagrams and the three-component aqueous diagrams. These systems are useful not only in the study of the systems but also in the search for methods to stabilize a product from either solid-solid phase transitions the adsorption of water from the atmosphere.

Phase chemistry may put a caking project on a firm scientific footing when the project might otherwise be little more than a search in hopes that some magic substance can be uncovered. The formation of solid solutions is a very useful tool when searching for a substance to lower the solubility of a component in a formulation.

REFERENCES

40. Ricci, J.E., *The Phase Rule and Heterogeneous Equilibrium*, D. Van Nostrand Company, New York (1951).

References / 113

41. Gleick, J., *Chaos*, Penguin Books, New York, N.Y. (1987); Bagnold, R.A., *The Physics of Blown Sand and Desert Dunes*, W.M. Morrow and Company, New York, N.Y. (1943).
42. Barndorff-Nielsen, O.E., *Acta Mechanical 64*, 1 (1986).
43. Ubbelohde, A.R., *Quarterly Rev.*, *11*, 246 (1957).
44. *Encyclopedia Of Science And Technology*, Vol. 12, p. 126, McGraw-Hill Book Company, New York (1960).
45. Macer, J.E., and Streeter, S.F., *J. Chem. Phys.*, *7*, 1019 (1939).
46. West, P.W., *Chemist-Analyst*, *43*, 1 (1954).
47. Griffith, E.J., *Chem. Eng. Data 8*, 22 (1963).
48. Patridge, E.P., Hicks, V., and Smith, G.V., *J. Am. Chem. Soc. 63*, 454 (1941).
49. Morey, G.W., and Ingerson, E., *Am. J. Sic. 242*, 4 (1944).
50. Hill, W.L., Faust, G.T., and Reynolds, D.S., *Am. J. Sic. 242*, 547 (1944).
51. Berak, J., and Zmanmievowska, T., *Roz. Chem. 46*, 1971 (1972); ibid. 1697 (1972).
52. Morey, G.W., *J. Am. Chem. Soc. 74*, 5783 (1952).
53. Gremier, J.C., Martin, C., and Durif, A., *Bull. Soc. Fr. Mineral. Cristallogr. 93*, 52 (1970).
54. Griffith, E.J., *Crystalline Calcium Polyphosphate Fibers*, Am. Chem. Society Symposium Series 171, 361 (1981).

CHAPTER

FIVE

Electrically Induced Cake Formation

ELECTRICAL CRYSTALS

Definitions

Piezoelectric crystals *are crystals that develop a charge when the crystal is mechanically stressed.*
Pyroelectric crystals *are crystals that develop a charge when the crystals are heated or cooled.*
Ferroelectric crystals *are crystals that become spontaneously polarized when cooled below a transition temperature. Polarization may be aligned or reversed in an electric field.*

The hierarchy of electrical crystals is piezoelectric crystals are the most plentiful and twenty of the thirty-two possible crystal point groups, or classes, exhibit piezoelectric behavior, to some extent [55]. It has been reported that twenty-one crystal types should exhibit piezoelectric behavior but one is too weak to consider [56]. Pyroelectric crystals follow in number with ten of the twenty piezoelectric exhibiting pyroelectric behavior. Lastly, the ferroelectric crystals depend upon the formation of ferroelectric domains similar to ferromagnetic domains for their behavior. Only crystals that have no center of symmetry can exhibit the piezoelectric behavior. It is also true that only crystals that lack a center of symmetry can exhibit pyroelectric, or ferroelectric behavior.

In passing it should be mentioned that there are 230 different kinds of crystals when space groups are considered. The space group adds glide planes and screw axes to the symmetry elements mentioned previously. There is no direct evidence that space groups may be related to caking problems but as time passes perhaps they will be of more use. At this time it is still not possible to predict whether or not a substance will or will not cake if armed with fundamental principles only.

Consider for a moment a long slender crystal that has been shown to be birefringent in the polarized light of a petrographic microscope; that is it has two or more indexes of refraction. If this crystal is stressed by twisting it, the positive and negative charges in the crystal will be displaced with respect to each other. This will cause one end of the crystal to develop a positive charge and the other end to develop a negative charge. If the torque on the crystal is reversed, the end that was negative will become positive and the other end will also have reversed charge. If electrodes are connected to the crystal, the crystal may behave as an electrical generator. If the charge on the crystal cannot be dissipated in some way, then the charged crystal will attract other charged crystals with the reverse charge configuration. This may be the onset of caking or it may contribute to a solid cake. It may also cause powders that are loaded and unloaded from cars by air conveying to be difficult to control.

If the crystals in a bag of piezoelectric crystals are stressed, and then the stress is relieved, the crystals in the bag will move ever closer to each other as the stress is applied and removed repeatedly. As the crystals approach each other closer and closer, the contact area will increase and any caking tendency from other causes is magnified by the electrical behavior even when the electrical behavior is not the primary cause of caking.

The conditions with pyroelectric and ferroelectric behavior are similar to the piezoelectric behavior except that the charge is created by heating or cooling in the case of the pyroelectric crystals and by electric fields in the case of the ferroelectric crystals. The ferroelectric crystals are a special case because they may polarize spontaneously and may remain polarized even after the electric field has been removed.

ELECTRICAL CAKING

When electrically induced caking is considered, more often than not, static charges come to mind. Too seldom are piezoelectric, pyroelectric, and ferroelectric properties even considered. The kind of electricity contributes to caking problems depends on the substance and its environment. Powdered substances that are conveyed pneumatically will be more inclined to suffer from static charge problems which substances subjected to pressure changes at the bottom of a stack or in the bottom of a loaded truck will be more likely to encounter the conditions which lead to piezoelectric effects.

Assume, for a moment, one has grown a batch of well-formed monoammonium orthophosphate crystals. Since these crystals grow very easily from saturated solutions, this is not a difficult task. The relative humidity over a saturated solution of $NH_4H_2PO_4$ is 93% and the low solubility is 35.7g per 100g of water at 20°C [57]. This is almost equal to the solubility of sodium chloride at the same temperature. Solubility per se is not a reason to expect that protected production of monoammonium orthophosphate to cake. Monoammonium orthophosphate forms no hydrates in the ambient temperature range. This is not a reason to expect the salt to cake. Also there are no phase transitions in the ambient temperature range. Transitions are not suspect. The crystals are well formed and there is little plastic flow and the shapes of the crystals are such that they should slide past each other with ease. Most of the reasons salts cake have been eliminated and yet monoammonium orthophosphate is known to be difficult to handle because *it cakes!*

There is small wonder that monoammonium orthophosphate cakes when one considers what happens to the crystals when the salt is placed at the bottom of a stack or in a bouncing freight car. Every crystal in the load is polarized each time it is stressed and the load begins to act as a group of electrostatic magnets. To make matters worse, the better the quality of the product the more difficult the caking problem becomes. If a non-caking form of ammonium orthophosphate is to be realized, it must have some way dissipate the charge on the crystals. Conduction of the charge to ground would probably work if the crystals were treated with a small quantity of a conductor on the surface of the crystals. A plant chemist recently reported that a rail car of monoammonium orthophosphate had arrived badly caked. In order to break up the load enough that it could be removed from the car, the operators selected a long iron bar. An operator reported to the chemist that he received an electrical shock when the iron rod penetrated the surface of the monoammonium orthophosphate.

MECHANISMS FOR PIEZOELECTRIC CAKING

Before leaving the ammonium orthophosphates some mention of the chemistry of the phosphates must be included because the caking is certainly not totally electrical. Both mono- and diammonium orthophosphates cake and the diammonium orthophosphate is considered to be a much worse example than the monoammonium orthophosphate by the plant personnel who have responsibility for both products. It is not surprising that the phosphates cake because the phosphates are not very stable and when the electric effects bring the surfaces in contact, any loss or gain of ammonia will cause the surfaces to grow together. Triammonium orthophosphate cannot be prepared and diammonium orthophosphate will usually have an odor of ammonia. The reasons for the unstable nature of the ammonium orthophosphates become apparent when it is recognized that the orthophosphoric acid is a relatively weak acid. Van Wazer concluded after an extensive study of the literature that the first hydrogen of H-H_2PO_4 has an ionization constant of $7.5 \cdot 10^{-3}$ and is much weaker than HCl. The second hydrogen H-HPO_4^- is much weaker yet having an ionization constant of $7.99 \cdot 10^{-8}$, which is weaker than citric acid, while the third hydrogen H-$PO_4^=$ is $4.8 \cdot 10^{-13}$, which is extremely weak and can be displaced only by high concentrations of bases or precipitating cationic functions [58]. It is easily understood why the ammonium ion is not strongly held in a crystal lattice with ionization constants of the values exhibited by phosphoric acid.

The particles in cakes formed from monoammonium phosphate can usually be scratched apart with a fingernail. The bridging or penetration of the particles is mostly surface contact. This behavior suggest that the cakes are formed because the stress exerted on the crystals can be partly relieved by the migration of ions across the interface of the crystals in contact. This would be expected to hold the crystals together but would not be expected to form cakes of the type which form from transitions or crystallization of solids from solution between particles.

If a car load of piezoelectric crystals has been stressed, each crystal will become an electric dipole. If the adjacent crystals are free to do so, they will become aligned + − + − + − + − + − + − throughout the car and this will tend to relieve some stress on the system. Charged crystals are deformed and the positive end of the crystal is positive because it is richer in cations than the negative end of the crystal, which is richer in anions. In systems composed of semiconductors, one could speak of holes and electrons and positively and negatively doped crystals. The theory is well established as to what happens at the interface between a positively doped semiconductor and a negatively doped semiconductor when the are brought into contact [59].

Negatively doped crystal contains an excess of electrons while the positively doped crystals contain positive holes. The excess electrons are usually supplied by an element such as phosphorus or arsenic while the positive holes are created by doping with a trivalent element such as boron. The electrons migrate rather freely as semiconductors because they result from the unused electron of phosphorus, which can have a valance of five as in PCl_5. The holes, on the other hand, migrate by the movement of covalent bonds in the body of the silicon and their path is more directed. When a $p-n$ junction is formed by placing a positive crystal in contact with negative crystal, thermal activity causes some of the holes to migrate into the negative crystals and some of the electrons to migrate into the positive crystals. This will continue until there is a depleted region at the junction, and migration will cease because of the reverse attraction. An analogous behavior is probably functioning in the caking of monoammonium orthophosphate and other piezoelectric crystals.

There is probably a migration of charged ions across the interface of contact and a depleted barrier zone is formed between the two crystals. The driving force is the charge on the crystals and it will be lessened by a migration of ions but the migration will cause the product to cake. In the case of the undoped monoammonium orthophosphate crystals, it is probably a cation NH_4^+ or H^+ that migrates across the junction rather than a hole. This should result in a small mass diffusion across the junction. Despite the very great tendency of monoammonium orthophosphate to cake, the cakes are not usually very strong.

As noted the charges on the piezoelectric crystals develop because the cations and anions are displaced in opposite directions as a stress is placed on the crystals. When two charged piezoelectric crystals come in contact, the charge may be relieved in a similar manner to the doped semiconductors except in this case it is ions that are migrating between the crystals rather than holes and electrons. This would tend to cause the crystals to deform back to the unstressed shape and when the stress is removed, the crystals would remain in contact by very evident but weak bonding. It is known that crystals of monoammonium orthophosphate, for example, tend to cake more in hot humid conditions and this should aid the caking conditions described above.

The following data were determined by George Cadwallader of the Monsanto Company to test the influence of an electrical conductor, such as graphite, on the surface of monoammonium orthophosphate when a pressure of eight pounds per square inch is applied to monoammonium orthophosphates crystals contained in a cylinder caking mold about 1.5 inches in diameter and 1.5 inches in depth. Platinum wire electrodes were inserted on opposite sides of the cake parallel to the walls of the polymethyl- methacrylate cylinder. The potential difference between the electrodes was determined at right angles to the load on the crystals as a function of time in a humidity cabinet controlled at 32°C and 70% relative humidity.

TABLE 3 The Potential Developed across a Cake of Monoammonium Phosphate at 32°C and 70% Relative Humidity under 8 lbs/in² Pressure

Cumulative Time Hrs.	Millivolts	Millivolts Leads Reversed
0.0	0.2	−0.2
1.75	1.4	−1.4
42.75	1.5	−1.8
67.50	1.3	−1.8
143.75	1.0	−1.7
163.00	1.0	−1.2
189.00	1.1	−1.3
212.25	1.0	−1.3
240.75	0.6	−1.0
330.75	0.8	−0.9
336.75	Remove sample. Retain weights. Cool to 23.7°C at 70% relative humidity.	
Time at 23.7°C		
35 min.	0.2	−0.2
55 min.	0.1	−0.2
1.25 hrs.	0.1	−0.1
19.25 hrs.	0.1	−0.1
19.25 hrs. Remove weights.	0.1	−0.1
19.92 hrs.	0.0	0.0

The increase in potential as the salt is heated is probably a pyroelectric influence. The reason for the maximum in the potential may be a result of ion migrations across the interfaces relieving the stress on the system as the particles cake. It is the result to be expected if indeed there is ion migration between the charged crystals. It is not understood why the reversed leads were consistently larger values than the initial measurements unless the system was polarized when the first measurements were made or that there was unexpected diode influence in the pressured system.

The pyroelectric behavior was much more pronounced when the monoammonium orthophosphate was coated with graphite. The reversal of potential was not expected. It is known that electrets reverse signs on aging and that the positive side becomes the negative side. Electrets are prepared by allowing certain liquid waxes to crystallize in a strong potential field. They can be rather strong electrostatic "magnets", attracting and repelling each other by electric fields rather than magnetic fields.

Under the conditions of the previous test both the control and the graphite treated monoammonium orthophosphate crystals caked. It will be interesting to attempt to devise an experiment to determine

TABLE 4 The Potential Developed across a Cake of Monoammonium Phosphate with 0.2% Graphite Powder Coat at 32°C and 70% Relative Humidity

	Under 8 lbs/in² Pressure	
Cumulative Time Hrs.	Millivolts	Millivolts Leads Reversed
0.0	0.1	−0.4
0.17	0.6	−0.4
2.03	2.8	−3.3
19.25	0.8	−0.5
45.25	0.6	−0.6
68.50	0.2	−0.4
97.00	−0.2	+0.4
187.00	−1.8	1.8
236.25	−1.2	1.9
335.75	−1.8	1.8
336.08 Sample returned to room temperature.		
Time At 23.7°C		
0.02	+2.8	−1.1
1.42	−0.3	+0.3
1.59 Weights removed.		
1.84	0.1	0.0

whether or not the crystals continue to cake when the charge is bled from the mass of crystals.

Pierre and Jacques Curie were the first to discover that piezoelectric crystals not only became charged electrically when they were distorted but that but that if an electric field was imposed across a piezoelectric crystal the crystal became distorted [60]. They showed that if two crystals were clamped together and an alternating current was imposed on one of the crystals an alternating current could be developed across the second crystal. Piezoelectric crystals are capable of vibrating at very precise frequencies when excited with an alternating current. As a consequence of this property, quartz crystals in particular are at the very heart of all types of electronic gear requiring precision oscillators. Such everyday equipment as digital watches, color televisions, computers, calculators, cash registers, and automobiles, not to mention a very great variety of test and measuring equipment depend upon this oscillating property of piezoelectric crystals.

Every crystal in a load of monoammonium orthophosphate, ammonium nitrate, ammonium sulfate, tartaric acid, or other piezoelectric crystals is capable of becoming a charged dipole. If this does not cause the solids to cake at a minimum, it causes the crystals to be difficult to unload if they have been stressed in transit in a bouncing truck or rail car.

A very interesting paper was published, which deals with the caking and flow characteristics of magnetic particles in a magnetic field. As might be expected magnetic particles will cake as a result of the magnetic fields that interact. This is similar to the influence one can expect from electric fields around electrically charged crystals. The influence is more or less permanent in ferroelectric crystals but can be exhibited by both piezoelectric and pyroelectric crystals as well. It was demonstrated in the case of the magnetic particles that uniform magnetic fields influenced the packing of magnetic particles. It was shown that the bulk densities of the 50 to 100 micron particles could be *decreased* by 10% to 20%. Interestingly it was also demonstrated that the hysteresis influence of the magnetic particles caused the results to vary as the field was applied at different stages of the packing process. It was concluded that the magnetic field influenced the packing in two ways. It caused the particles to stick together but at the same time the particles were more loosely packed [61].

PYROELECTRIC CRYSTALS

Pyroelectric behavior is similar to piezoelectric behavior except that it adds a new dimension to the problem of caking. As mentioned, any crystal that is pyroelectric is also piezoelectric. The added dimension is that the crystals develop an electrical dipole when the crystals are being heated or cooled. When the crystals are heated, one end of the crystals become positively charged while the other end becomes negatively charged. When the same crystals are cooled, the end that was positive now becomes negative. Obviously as a product of this type is heated in a warehouse or rail car, the crystals move to form a tighter bonding. One obvious point to check in a caking problem of any product containing a pyroelectric crystal is whether or not the product is being loaded for shipment when it is still hot.

SEED CRYSTALS

One of Kundt's test for pyroelectric crystals was to exhibit vividly the electric field surrounding a charged pyroelectric crystal [56]. The most impressive demonstration is obtained by allowing naturally highly charged MgO smoke obtained by burning a magnesium ribbon in a closed vessel covering a heated pyroelectric crystal to make the field visible much as iron fillings can be used to make a magnetic field visible. Electric fields of this type are certain to have an influence when these crystals are used as seed crystals in melts and solutions.

Kundt's second test for pyroelectricity is also spectacular but does not demonstrate that the field extends well beyond the boundaries of the crystal faces themselves. In his second test red lead and sulfur are blown on to the charged crystals with a blow gun. The crystals emerge from the test red on one end and yellow on the other if the test is positive. This definitely shows that the crystal is polar and that a charge exists on the crystals but the field is implied rather than demonstrated. It is doubtful that the field exists around the crystal after the oppositely charged powdered particles stick to the crystal. As suggested, this is probably action similar to that of many flow conditioners.

The phosphate and silicate systems are referred to as glass formers and they are highly susceptible to seeding. An example is sodium Kurrol's Salt, $[NaPO_3]_n$, where n is a very large number, 10,000 or more. It is composed of very long helical chain molecule-ions and has a crystal property remarkably similar to asbestos [58]. It is, therefore, a polyphosphate with a metaphosphate composition and is metastable in the Na_2O-P_2O_5 phase diagram. Properly seeded sodium phosphate melts can grow into sodium Kurrol's Salt but without seeds the preparation is hit or miss.

It is interesting to speculate as to whether it is the electric field around a crystal that allows it to be a seed crystal or whether the crystal is acting as a templet for additional growth. There is some similarity between the DNA molecules and the sodium Kurrol's Salt molecules. It may be suggested that the seed may act as more than just a templet. When a melt is seeded with an acicular or fibrous crystal that floats horizontally on the surface of the melt, there develops a region around the seed crystal that is similar in shape and structure to the electric field about the pyroelectric crystal that has been dusted with MgO smoke.

The action of seed crystals is not limited to melts and the action is equally interesting for growing crystals from solution. Extensive unpublished work has been done with the supersaturated solutions of Form II $Na_5P_3O_{10}$. As mentioned often, $Na_5P_3O_{10}$ cannot have a solubility in the strictest meaning of the concept because it is degraded in aqueous solutions. The hydrolytic degradation is slow enough to allow a reasonable value to be obtained at lower temperatures and intermediate pH values. The system approaches but never attains an equilibrium that is drifting ever so slowly toward lower pH values and ultimately orthophosphates. The solid phase in the supersaturated solution will be $Na_5P_3O_{10} \cdot 6H_2O$ when crystallization begins but seeding the system with crystals of $Na_5P_3O_{10} \cdot 6H_2O$

had little influence in causing the supersaturated solutions to begin to crystallize. Crystallization occurs but the rate is independent of the presence of $Na_5P_3O_{10} \cdot 6H_2O$ crystals. This behavior is unlike supersaturated solutions of sodium acetate, which may instantaneously become almost solid when seeded with a small crystal of solid sodium acetate.

Seed crystals can have a very great influence on the caking of solids particularly if they are causing a transition to occur that might otherwise not happen or if they are triggering a transition at a lower temperature than it might otherwise happen. The seed crystals may not seem to be related to the product in any way and yet they may cause problems to appear. Multi-product plants may cause problems where more than one product is made in the same equipment. It is known, for example, that trace quantities of potassium ions will trigger the transition of Form II $Na_5P_3O_{10}$ to Form I $Na_5P_3O_{10}$ at lower temperatures while small quantities of acid magenta dye will inhibit the Form IV–Form III transition of ammonium nitrate merely by coating the surface of the ammonium nitrate. When caking problems occur under conditions that may not be predicted by the known chemistry of the system, seed crystals are at least worthy of consideration. In most cases problems would likely be caused by solid–solid transitions.

FERROELECTRIC CRYSTALS

Ammonium nitrate exhibits pyroelectric caking but it exhibits all of the electrical and chemical forms of caking except hydrate formation. It forms no hydrates. Monoammonium orthophosphate is also pyroelectric. Relatively few crystals exhibit ferroelectric behavior, but monoammonium orthophosphate has it all and ammonium nitrate is not far behind. The ammonium orthophosphates and ammonium nitrate are excellent pilot substances to study while either learning the techniques used to study caking or for demonstrating the properties to others.

In the 1920's and 30's some very interesting work was published dealing with the electrical properties of ferroelectric crystals and one of the more studied substances was Rochelle salt, sodium potassium tartrate tetrahydrate, $NaKC_4H_4O_6 \cdot 4H_2O$. It is not too surprising that much difficulty was encountered in understanding the strange behavior of this salt. Even today it is difficult to get a really crisp definition of ferroelectricity. Using equipment such as rheostats and ballistic galvanometers, the workers recognized that the crystals were not behaving like usually encountered piezoelectric crystals. The die-out of charge induced on the crystals was charge dependent and reversing the electrodes consistently gave different results. It was soon recognized that a permanent polarization was leading to hysteresis in the charge and discharge of the crystals. It was also recognized that crystals of this type exhibited remarkable dielectric constants in certain temperature ranges [62] [63] [64]. The work was fantastic in its simple beauty and is surely worthy of an evening's study [65].

It is interesting to note that very large crystals may be very easily grown from saturated solutions of ferroelectric crystals. Ammonium orthophosphates, monopotassium orthophosphates, and sodium potassium tartrate tetrahydrate have all been grown from aqueous solution to single crystals weighing thirty or more pounds. With sufficient care most salts could be grown to single crystals this size but the ferroelectric crystals are especially easy to grow.

The dielectric constant of a crystal is related to its refractive index. For reasons that are not well understood titanium salts have very high refractive indexes and for this, reason titanium dioxide is used in many applications where hiding or covering power is desired. Paints containing TiO_2 cover better than any others known. On the other hand the refractive index and the dielectric constant of crystals such as barium titanate is also unusually large. This relation is to be expected from the Clausius-Mosotti equation, which deals with molar polarization, and the Lorentz-Lorenz equation, which deals with molar refraction. Both molar polarization and molar refraction are a form of molar volume and at very long wave lengths of light it may be shown in theory that

$$D = (n_e)^2 \qquad [17]$$

where:
    D is the dielectric constant and
    n_e is the refractive index

It may be expected that if a crystal has a high refractive index as determined easily by the optical microscope it will probably have a high dielectric constant also. Crystals of this type may be inclined to flow poorly particularly if they have been exposed to electric fields.

Although there are fewer ferroelectric crystal than pyroelectric crystals, they are an extremely interesting class of crystals because of their many uses. It is well known that crystals of the perovskite type (high temperature form is simple cubic, eg. XQO_3 or $KNbO_3$, $AgNbO_3$ and $PbZrO_3$) are more likely to exhibit ferroelectric behavior. A ferroelectric crystal is a crystal that becomes polarized when it is placed in an electric field in much the same way that a ferromagnetic substance becomes magnetized in a magnetic field. In the magnetic

Ferroelectric Crystals / 125

field the ferromagnetic substance develops magnetic domains, which are easily seen with a microscope. As the strength of the magnetic field is increased, the size of the magnetic domains is increased at the expense of their number. This may also be seen with a microscope. When the magnetic field is reversed with respect to the magnetic crystal, the direction of polarization of the crystal is also reverses unless the crystal is free to rotate in the magnetic field.

The analogy between the ferromagnetic crystals and the ferroelectric crystals is remarkable. In the previous paragraph, if electric field were substituted for magnetic field and ferroelectric crystal were substituted for ferromagnetic crystal, the exact wording could be repeated including the formation of electric domains and their visibility in electric fields. When the electric field is reversed, the direction of polarization in the ferroelectric crystal is also reversed. If the "poling" field is strong enough, the ferroelectric is left with electric polarization in just the same manner that a magnetic crystal is left with magnetic polarization. In other words, the ferroelectric crystal can behave as an electric "magnet" in the sense that the crystals exhibit poles and are attracted and repelled as magnets are attracted and repelled.

The reason a ferroelectric crystal behaves as it does is relatively easily understood after the fact when it is known that a substance is ferroelectric in its behavior. Some of the ions in a ferroelectric crystal have two or more equilibrium positions in the crystal, but the equilibrium positions are displaced from the center of the lattice cells that make up the crystal. It may be thought of as a hill in the center of the cell with walls so steep the ions cannot abide there. If an electric field is imposed in one direction, the ions are displaced in that direction, and if the electric field is imposed in the other direction, the ions are displaced in that direction. But the ions cannot inhabit a region of the cell that does not contribute to the polarization of the crystal. The more cells that are polarized in the same direction the stronger the polarization of the domains. When the electric field is removed, the crystals will retain some of the polarization in the same way that the ferromagnetic crystals retain magnetism. The residual polarization of the crystals may contribute to the caking and lumping of these salts.

In considering potential methods of solving caking problems caused by electrical caking, the options are many but each system will dictate whether or not an option is practical. The trick is to either dissipate the charge by conducting it away or to isolate it in a manner that will not allow dipoles to interact. As an example the pyroelectric crystals, which were treated with the read lead and sulfur, should remain free flowing at least for a few cycles. The red lead and sulfur is a part of Kundt's second test for pyroelectricity and is discussed in detail in Chapter VI. In this test the charges at the ends of the crystals should be neutralized by colloids of the opposite charge to the polarized crystals. Surely most systems could be treated with charged colloids, which are colorless. In the case of ammonium nitrate, magnesium oxide worked very well indeed but there was some initial chemical reaction with the surface of the particles. Also any type of ionizing radiation will reduce the charge on the crystals and in the case of the calcium carbonate, limestone, mentioned earlier, the oleic acid acted as an insulating dielectric preventing the particles from becoming charged.

Again this discussion is intended to be only introductory. An in-depth coverage would require many volumes and one could easily spend a productive lifetime in this and similar studies. It is the desire of the author to give the reader a method of determining whether or not a caking problem is caused by electrical behavior and to suggest a starting approach to a solution to the problem. It should be admitted that in certain products there is no solution that may be tolerated. Markets that require ultra pure chemicals may be of this type, and one copes with caking in order to obtain other necessary properties. A note of advice; *do not assume that a treatment cannot be tolerated without first checking with the customer.* Many times that, which is at first believed to be unacceptable, may be very desirable after the fact. This is particularly true once it has been demonstrated that a treatment is very effective.

STATIC ELECTRICITY

Static charges may cause solid powders to misbehave in many ways. Anyone attempting to weigh light powders on a very dry winter's day has experienced the frustration of touching the powder only to have it fly all over the balance as if it had exploded. This happens when the powder becomes charged with an excess of either positive or negative charge. The example of a static dipole was discussed with respect to bridging with flow potentials and the soft cakes that can form from bag set or bottle set. Please see Figure 22. These may be detected by shaking the particles vigorously to break any lumps and then allowing the material to sit undisturbed for fifteen or twenty minutes. If static charge was responsible for the lumps, they shall have returned to a greater or less degree within this time frame. An electrometer, tritium gun, or charged wand may also be used to advantage in this test. One of the more interesting tests consists of suspending a crystal between two plates, which have been attached to a high voltage DC power supply. When the crystal contacts one of the plates, it will become charged, thus being attracted to the other

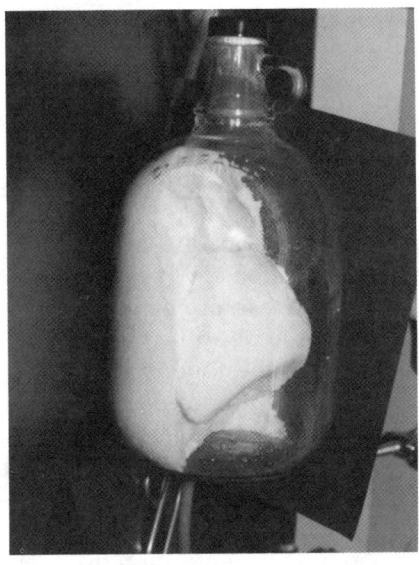

FIGURE 22. *A lump formed in prilled ammonium nitrate caused by electric charge. The lump will form repeatedly after the ammonium nitrate has been broken up into a free flowing system.*

plate. When it contacts the other plate, the process is reversed and the crystal will oscillate between the plates as long as the power is supplied. Most dry crystals will respond to this test.

Charges on Crystals

Electrostatic caking is seldom as strong as crystal dipole caking but it may become a serious problem particularly when solids are conveyed in pipes. A solid can build up so much flow potential in a pipe as to create bridges in pipes of large diameter and these bridges can support hundreds of pounds of powder. The late Dr. B. B. McCann, formerly with the Calcium Carbonate Company of America, told of a case study that vividly illustrated the problems that may be caused by electrostatic charges on inert solids.

A well-known St. Louis-based feed company mixed limestone in its feeds. Limestone is about as inert as chemicals are usually found. The company's feed mixing room was in the basement of a seven-story building and the railroad unloading platform was on a bluff at the top of the building. The company conveyed the finely powdered limestone to the basement through a series of metal pipes of fairly large diameter. The flow of the white powder was controlled by a series of dampers in the pipes. From time to time the finely powdered limestone would build up a static charge on its surface great enough to cause the limestone to bridge the pipe halting the flow. Usually tapping the pipe sharply for a few times was sufficient to cause the flow of the powder to begin.

One day a vice president of the company was touring the plant with a distinguished guest. When the group reached the basement mixing room, they gathered around to see the limestone delivered through the pipes. At this instant a bridge of electrostatically held limestone broke free and dumped several hundred pounds of white powder on the guest and the vice president causing the vice president to become very angry. He ordered that this problem be eliminated at once. The company approach Dr. McCann. He informed them that he felt that he could cause the powered limestone to be free flowing.

Dr. McCann supplied a bulk carload limestone that had been ground in water containing a very small quantity of oleic acid. The carload was unloaded into to bin at the top of the building. Now that the static charge, which had previously allowed the powder to be controlled by dampers, had been eliminated, the powder flowed as a fluid. Once the flow of the limestone was started, there was no stopping it with the dampers no matter how desperate the effort. The powder flowed though all of the dampers in the pipes and the entire carload was delivered to the basement mixing room as one uncontrollable stream.

There are three lessons to be gained from the previous account. First, controlled caking is not necessarily bad. Second, static charge may be eliminated on selected inert solids. Third, drastic changes in the properties of a product should be tested on less than carload quantities before the product is delivered to a customer.

An additional warning should be given concerning the magnitude of the flow potentials that may be built up on transported particles, particularly as they become more and more charged in equipment that is not properly grounded. The electrical discharge is great enough at times to cause fire or explosions in dust collector bags that are not properly grounded [66]. This is not surprising if one considers that lightening is nothing more than the discharge of the flow potential from colloidal mist droplets. Johnston and White report that on alumina, Al_2O_3, potentials as large as 5,000 volts may be generated causing problems in transporting the powder [67].

Static charges on polymer powder feed stock and on textiles has been troublesome for many years. It has been reported by some textile workers that on some nights it becomes difficult to keep cloth folded as it exits the machines. Schwaegerle reports that with PVC powders grain shape is mainly responsible for bulk density of the powders while grain size and the distribution of particle size are primary variables in powder flow but that static charge may overwhelm all other influences [68]. Masui studied the build up of charge on polymers as they were passed through pipes [69]. He concluded that the charge on Nylon 12 was acquired mainly at U bends in the piping but had little effect on high density polyethylene. O'Neill and Willis injected charge with a corona discharge into pneumatically conveyed powders of the type used in the electrostatic coatings industry [70]. They found that the charge injected on to the powder could be detected and it could be determined what the rate of flow was in a limited way. It was also noted that the powders would seek the walls just after being charged but would reenter the stream of powder in a short time.

Ivanov and Isaev developed equations calculating the electrostatic components of pressed particles as in pelletizing [71]. They concluded that the dissolved salts between the particles in any solution influenced the force acting on the particles and was inversely proportional to the concentration of the dissolved salt in the solution coating the particles.

Positive and Negative Charges

An unpublished study was done on the attraction of particulate to surfaces in an attempt to better understand how a detergent is required to overcome forces when it became desirable to demonstrate whether or not there is a fundamental difference between a positively charged surface and a negatively charged surface other than the attraction-repulsion of pith balls, electrometers, and so forth. It was also desirable to demonstrate the differences in a direct, simple way if the fundamental difference existed. It had been recognized that positive and negative charges did not distribute themselves in the same manner over a surface and that most of the movement was associated with the negative electrons.

The following description demonstrates the differences in structure of positive and negative fields of static electricity. Very few supplies and equipment are required for the demonstration. These materials are needed: one pad of writing paper, 8.5 inch × 11 inch; one sheet of cellulose acetate, 8.5 inch × 11 inch; one sheet of polyvinyl chloride or polyester, 8.5 inch × 11 inch; a small bottle of powdered activated charcoal; a box of tissue paper; and a spatula.

Attach a sheet of cellulose acetate to a sheet of polyvinyl chloride. Lift one sheet of paper on the writing pad and place the sheets to be charged on top of the writing pad, below. Close the top sheet of paper on the transparent sheets. Now, using a few sheets of tissue rub the top sheet briskly for about two minutes. Remove the plastic sheets from the writing pad. Separate the sheets at a corner and pull the sheets apart with a smooth continuous motion. When the sheets

130 / Electrically Induced Cake Formation

are separated, the cellulose acetate is usually positively charged while the polyvinyl chloride is charged negatively. It is usually best to hang the sheets upon a hook until they are developed. Care should be taken to prevent the sheets from touching because they will attract each other.

The development of the sheets is very easily done. While holding a sheet vertically over several sheets of tissue used to catch any falling charcoal powder, place about two grams of charcoal on a spatula. Hold the spatula about two inches above the top edge of the sheet and about two inches away from the surface. Slowly sprinkle the charcoal from the spatulas allowing the charge of the sheet to attract the falling charcoal. With a little practice the sheets may be developed to be as dark as one may choose.

Several characteristics of statically charged bodies become obvious. The tree shape of the negatively charged image is very similar to lightening in outline. The branching structure suggest some kind

FIGURE 23. *Charcoal imaging of positive static electricity.*

of movement took place as a final placement of charge when the sheets of plastic were separated. Conversely, the positively charged sheet exhibits a cloud formation of less mobile "holes", which were formed when the electrons were removed. It is unlikely that cations, per se, contribute much to the positive charge on a sheet of cellulose acetate.

Many other surfaces that become charged may also be developed. Charged black cloth may be developed with magnesium oxide powder. The technique is similar to the development of fingerprints with finger print powders. It is interesting to find that both positive and negative fields may exist simultaneously on a nonconductive, charged body. Therefore, a spot on a garment may often be removed by rubbing it with another part of the same garment.

The behavior exhibited by the positive and negative charges may be observed in many other systems. Most solids capable of ioniz-

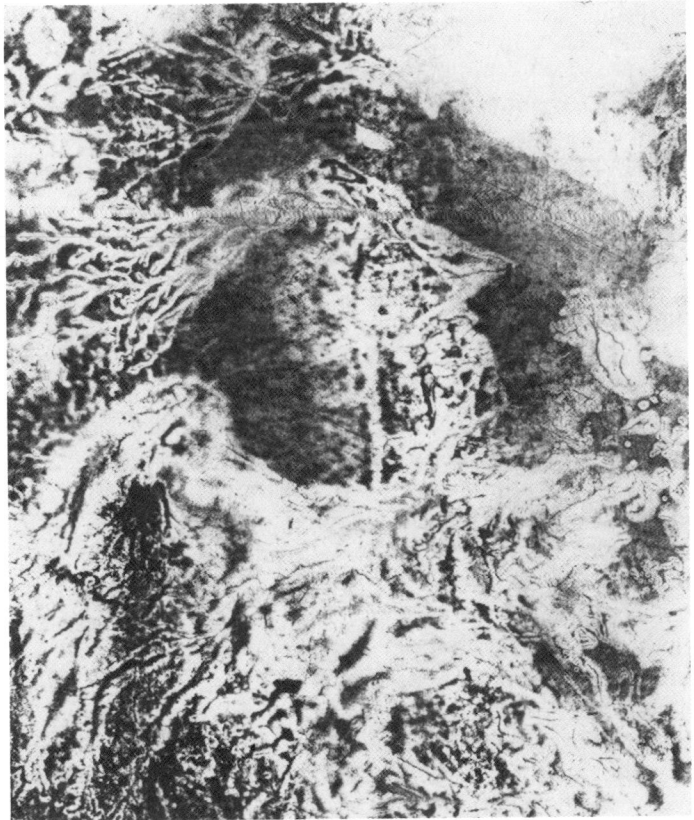

FIGURE 24. *Charcoal imaging of negative static electricity.*

ing in water will become negatively charged because the cationic function is more mobile than the larger anions, which make up the structures of clays, sands and similar solids. In the case of the organic molecules mentioned previously, it is necessary for the electrons to be removed from one structure and deposited on the second polymer. The negative particles are then the more mobile, and the positive charges must be left in the molecules that correspond to the anions of the inorganic solids. It is, therefore, reasonable to assume that the positive charge would be cloud shaped and the mobile electrons should leave a branching path. In all cases to which the test was applied, the results were the same. The sign of the charge was confirmed with the aid of an electronic electrometer.

Examples of the developed charges are shown in Figures 23 and 24.

FLOW CONDITIONERS

Flow conditioners are any substances that cause powders to flow more freely. They may be anti-cake agents as well. In general flow conditioners may be divided into two classes. The first class is usually derived from organic amines, alcohols, acids, and salts. Many have surfactant properties as well as coating a product with a lubricating surface that has a low dielectric constant and is probably a poor conductor of electricity. They form a barrier about the particles and will not allow a raw crystal surfaces to contact each other. The second type of flow conditioner is usually a fine-powdered, charged solid that may or may not be insoluble in water but most of the commercial products are insoluble. Typical solids are tricalcium orthophosphate, fumed silica, clays, magnesium carbonate, and soft burned magnesium oxide. The grade of magnesium oxide is extremely important if it is to be used as a flow conditioner. Only the soft burned, very active MgO is worthy of consideration.

To be effective it is necessary that the flow conditioner stick to the surface of the particles it is designed to protect. This is particularly true of the clay type conditioners that function to absorb any liquids that form on the surfaces of the product, and they work to prevent the surfaces of the crystals from coming into contact. The clays are used in conjunction with other conditioners on some products. The clays most often used are attapulgite (an acicular crystal), kaolin, talc, and bentonite. Price is often a major consideration when selecting a clay, but the type of product is also a consideration. Fertilizers are good candidates for clay treatments provided that the fertilizers are not to be used in applications as drip irrigation where it is necessary for the dissolved fertilizers to pass through the fine orifices of a spray nozzle. Pharmaceuticals are more likely to require either kaolin or bentonite because of their texture, color, and safety record. Many food products use tricalcium orthophosphate or fumed silica in a similar manner but they may included for their thickening properties as well.

Reactive conditioners have not been used to a great extent but the concept is open to investigation. One example is the coating of monoammonium orthophosphate with reactive MgO to form magnesium ammonium phosphate on the surface of the monoammonium orthophosphate. Magnesium oxide also works well on ammonium nitrate, which will form the magnesium nitrate-ammonium nitrate double salt on the surface of the particles.

Some very interesting experiments were performed by Nash, Leiter, and Johnson in which they measured the electrical charge of Carbowax 6000 and Cab-O-Sil and how the Cab-O-Sil influences the charge on the Carbowax [72]. They arranged two parallel electrode plates so that a powder could be directed as a stream between them. The electrodes were connected to a DC power supply in such a manner that one electrode was charged +7500 volts while the other was charged −7500 volts with respect to ground. It was learned that the powdered Carbowax 6000 was attracted toward the negative electrode while the Cab-O-Sil was strongly attracted toward the positive electrode. If 1% Cab-O-Sil was dusted on to the Carbowax, the two oppositely charged powders completely killed the charge on each other and the mixture was attracted to neither electrode and poured much like a liquid.

The charge test of Nash, Leiter, and Johnson utilizing the Carbowax system, although specifically designed for the Cab-O-Sil, should be very useful for optimizing the use of flow conditioners on charged products. There is little doubt that most solid flow conditioners behave by a similar mechanism. They not only kill surface charge but also place a contact barrier between the particles, a behavior not unlike that of a strip of adhesive tape when one attempts to affix it to a dusty surface. There is seldom attraction between the tape and the surface on which it sticks while the dust sticks to the tape. The collection of fingerprints and the attraction of fingerprint powder depends upon these same properties of tape, powders, and charged surfaces.

Charges on Liquids

The behavior of charges on liquids may be important to the main theme. If one measures the surface tension of a liquid, for example, an aqueous solution, with a tensiometer of the ring type, one will discover that a charge on the surface of the liquid will increase the surface tension of the liquid. The so-called mercury heart is a vivid example. A few drops of mercury are placed in a watch glass and then covered with a solution of six normal sulfuric acid containing a small quantity of potassium permanganate. A small iron nail is placed on the edge of the watch glass so that it will just touch the drop of mercury. When contact is made the mercury develops, a negative charge from the nail electrode and forms a tight ball withdrawing from the nail. The charge has increased the surface tension of the mercury. Next the potassium permanganate oxidizes the charge from the mercury and it relaxes to again touch the iron nail and the process is repeated continuously until the permanganate, iron, or acid is consumed.

Several interesting observations may be made about the influence of charge on liquids and how it will influence the behavior of the liquid. If a nonconductive dish contains a liquid capable of being charged when it is poured (most if not all liquids and solutions) from one container to another when first poured, the surface tension of the liquid will be above equilibrium if the depth of the liquid is more than about two centimeters. If, on the other hand, the depth of the measured solution is one or two millimeters, the surface tension will come to equilibrium almost instantly. The more vigorously the liquid or solution is shaken the greater the surface tension will be when the liquid is first measured. It make little difference whether or not the solutions are conductive of electricity provided the container from which they are poured is nonconductive.

As mentioned previously, the greater the depth of the measured liquid the higher the surface tension of the solution will measure and the longer the time required for the solution to acquire the equilibrium value. If one chose, the charge on the liquid can be transferred to a small sheet of Mylar, which has been grounded to remove any charge, by merely lowering a sheet of the Mylar perpendicularly through the surface of the liquid and removing it at once. It will be noted that the surface tension of the measured solution can be reduced several dyne. Moreover if the charged Mylar is then passed through the surface of a solution that has already reached the uncharged surface tension, the surface tension of the second solution will increase almost as much as the first solution decreased. It is necessary that the measured solutions are several centimeters deep when measured.

Yet another interesting experiment is to pass a charged wand over a charged solution. Fill a beaker, of almost any size, with a solution. A sodium chloride in water solution will serve. Now charge a wand. A one centimeter diameter solid glass rod about 30 to 40 centimeter will work well. Charge the wand by rubbing it briskly with soft dry tissue paper until it has become highly charged. Now pass the charged wand horizontally and rather quickly, about two cm. over surface of the liquid. A considerable quantity of the liquid will follow the wand and spill out of the beaker. It is probable that a similar condition contributes to the wave action at sea. A highly-charged cloud will pass near the surface of the sea attracting the water and contributing to wave action until the potential difference becomes large enough for lightening to jump between the water and the cloud.

SUMMARY

Electric charges can influence caking properties in a variety of ways. There are three types of crystals, piezoelectric, pyroelectric, and ferroelectric that have the ability to generate electric fields when stressed, heated, or poled with an electric charge. This property can lead to caking and handling problems in loading and unloading products. Little is known about the how these problems can be solved and much work is needed to better understand how to overcome them.

In addition to the innate properties of crystalline solids to become charged, there is also the matter of static electricity on crystals and amorphous solids as well. Here again flow problems can cause problems and some substances can even become internally charged (as electrets) causing the products to have flow problems.

Throughout this chapter a highly refined, very involved and extremely useful science has been treated in a casual way. Some of the properties influence caking in an oblique way but are very necessary in understanding a system that has formed a cake. In the use of ammonium orthophosphate in fertilizers, for example, few if any customers should care if the crystals of monoammonium orthophosphate deform when an electrical field is imposed across the crystals, but they are concerned if the ammonium orthophosphate is badly caked.

REFERENCES

55. Phillips, F.C., *An Introduction to Crystallography*, Longmans, Green and Company, London, p. 103 (1957).
56. Buerger, M.J., *Elements of Crystallography*, p. 188, John Wiley and Sons, Inc., New York, N.Y. (1956).
57. Sidell, A. and Linke, W.F., *Solubilities of Inorganic and Organic Compounds*, D. Van Nostrand Company, Inc., New York, N.Y., p. 379 (1952).

58. Van Wazer, J.R., *Phosphorus and its Compounds,* p. 481, John Wiley and Sons, New York, N.Y. (1958); ibid. p. 665.
59. Van Der Ziel, A., *Solid State Physical Electronics,* Prentice-Hall, Inc., Englewood Cliffs, N.J., p. 498, (1957).
60. Curie, J. and Curie, P., *C.R. Acad. Sc. Paris 91,* 294 (1880).
61. Jones, T.B., *Powder Technology 56,* 31 (1988).
62. Kittel, C., *Introduction to Solid State Physics,* John Wiley and Sons, New York, N.Y., p. 182 (1953).
63. Kittel, C., *Elementary Solid State Physics,* John Wiley and Sons, New York, N.Y., p. 90 (1962).
64. Von Hippel, A., *Handbook of Physics,* Ed., Condon, E.U. and Odishaw, H., McGraw-Hill Book Company, New York, N.Y., p. 4–103 (1958).
65. Valasek, J., *Phys. Rev. 17,* 475 (1921).; ibid. *19,* 488 (1922); *20,* 639 (1922); *24,* 560 (1924). Frane, J.G., *Phys. Rev. 21,* 348 (1923). Isley, F.C., *Phys. Rev. 24,* 569 (1924). Sawer, C.B., and Tower, C.H., *Phys. Rev 35,* 269 (1930).
66. Dahn, C.J., *Proc. Powders and Bulk Solids Handling and Processing,* Canners Exposition Group, Des Plaines, IL, p. 196 (1983).
67. Johnston, T.J. and White, R.D., *Metallurgical Soc. AMIE,* 315 (1985).
68. Schwaegerle, P.R., *J. Vinyl Tech. 7,* 16 (1985).
69. Masui, N., *Jpn. J. Appl. Phys. 23,* 1492 (1984).
70. O'Neill, B.C. and Willis, C.A., *J. Electrostatics 17,* 99 (1985).
71. Ivanov, S.V. and Isaev, E.A., *Izves, Akad. Nauk 1,* 19 (1986).
72. Nash, J.H., Leiter, G.G., and Johnson, A.P., *Ind. Eng. Chem. 4,* 140 (1965).

CHAPTER

SIX

Laboratory Techniques and Test Procedures

FOR MANAGERS

Aside from the hardware, theories, experiments, proofs, solutions, and reports there is the problem of managing a laboratory engaged in solving caking problems. At best, the solution to a caking problem is likely to be an improvement on an old product but is not very likely to result in the creation of a new product, although this does happen sometimes. It may even be difficult to prove that because a caking problem no longer exists that the product is more profitable. There can be little doubt that a free flowing product has a commercial advantage over one that lumps and cakes. The advantage to having a free flowing product may be more in keeping current customers rather than acquiring new ones.

The manager is faced with the problem of funding product research. How much money should be spent on a caking problem? What type of personnel should be assigned to a caking problem? How much authority should be delegated to laboratory personnel, plant operators? How much time should be reasonable to expect a solution? What proof of the solution should be required? What factors will cause the product to fail again? Can these factors be controlled in a reasonable way? Is the problem one of basic science, process controls, raw materials, packaging, transportation, or customer abuse? Is the caking problem obvious or is a more subtle, as rail cars that take an excessive time to unload because the product does not flow as expected? Can the customer misuse the product and should the customer be educated in ways to prevent the product from caking? These

138 / **Laboratory Techniques and Test Procedures**

questions will be addressed in the Introduction because it is unnecessary for managers to know the details of the laboratory procedures and practices discussed in the body of the book.

LABORATORY PERSONNEL

Although the concept of caking is simple and most persons have a reasonably good idea of the phenomenon, this does not mean that a solution to a caking problems is likely to be simple. The problems usually require an in-depth knowledge of basic chemical and physical concepts and for most workers a B.S. degree training is usually required unless one has exceptional natural abilities, which are rare but of great value when available.

Too often the discipline of training is ignored in assigning personnel to a project. A project may be assigned to a laboratory chemist because the chemist happens to have be available at that time whether or not the chemist has the proper training to understand the problem and what is required to solve it. A caking problem is much more in line with the training of a physical chemist, an inorganic chemist, or even an analytical chemist rather than a biochemist or organic chemist. It does not mean that the other chemists or engineers are any less capable but that their interest is not likely to be as great, unless it is their product that has the caking problem.

PROJECT DURATION

The time devoted to a caking problem will depend upon many factors other than the difficulty of the problem. Many important questions must be asked. "How much is the product worth?" "What is the expected life of the product with and without caking problems?" "What are the restrictions which must be imposed on efforts to eliminate caking?" "Can additives be used and if so how much?" "How much is known about the chemistry and physical properties of the product, before the project begins?" "Has adequate library work been permitted or required?" "Have company reports dealing with this very problem been written in the past?" "If so, have the reports been read?" "Has Management read the reports?" "Does your company have patents in this area?" There are cases where companies are unaware of their own procedures in their own patent literature!

Too often a manager may be inclined to feel that unless there is great visible activity nothing worthwhile is happening. No one can appear busy while learning in a library. Too often a few hours spent in library research can prevent weeks of useless work done at the bench. As a result too often busy hands are supported by empty heads. It is very easy to convince one's self that washing beakers is of significant value to the company when in fact it might be less expensive to destroy the beakers and get on with the more difficult head work.

As a rule of thumb, between ten and twenty percent of a worker's time should be spent on library research if that worker is expected to perform as a professional and to solve difficult problems within a reasonable time. If the worker is not capable of utilizing this much time profitably for the benefit of the company, the problem may lie with management and not the laboratory personnel.

As an average time, about three man-months should be required for an obvious problem with an obvious solution. One man-year is not an unreasonable time for a difficult problem without obvious solutions. In some cases there will be problems that have no satisfactory solution and problems requiring more than one man-year are likely to be of this type.

LABORATORY VERSUS PLANT

Where to do the experimental work is an important question and is critical to the success of a caking problem. Certainly if there is a caking problem, there is a plant-size operation making the product. This usually means that there is a choice in choosing samples for study. They can be made in the laboratory under controlled conditions or they may be a part of an Experimental Designed experiment in a plant. It is probably best to define the problem in the plants and then do research to solve the problem in the laboratory. Then the laboratory solution should be tested in the plant to determine whether or not the full-scale product responded as expected. Very often an operating plant is capable of making much better products than are available from laboratory preparations. This is particularly true when the product is a solid. The surface-to-volume ratio is usually much greater in laboratory equipment than in plant production units and the chances for contamination from the containers are higher in the laboratory.

Experimental Design and statistics are powerful tools in plant operations and should be used, particularly in the definition of problems. This does not mean statistics should be used in place of classical science but should be used to supplement the classical approaches. This brings up the age old question of when research should be terminated and when development should be started. Development is usually concerned with pilot plant or even full plant operations. It can be very expensive to experiment on a plant scale. Usually research is concerned with small scale samples and a minimum of manpower. Development and statistically designed experiments can involve an entire plant and the personnel to run the plant. It was the early intentions to avoid the statistical approach entirely in this book, not because it is not very useful, but because it may not be possible to do justice to the subject. In the chapter dealing with Laboratory Techniques an attempt will be made to at least introduce the subject and how it was used to help solve an industrial caking problem.

A problem has been adequately solved only when the customer feels that the problem has been solved. The customer usually feels that a problem has not been solved when the competition can supply a product which never cakes. If a product is almost as good as the competition's, a customer may continue to use the product, although not completely pleased with it. Ultimately the success of any project designed to end a caking problem depends on the critical eye of the customer. The final test is the reaction of the customer. Learn whether or not the customer is pleased. The customer may be internal to your organization or the client of a field salesman. Quality products know no artificial boundaries! What does the customer think should be done to improve the product even more? Is the customer having any problems with your product? Do your other customers recognize the same needs? The project has been a success when a follow through with customers confirms that there are no caking problems with your product.

REPORTS

Some organizations are sticklers for reports and demand that the reports be of "total quality", while others consider reports to be an onus and a waste of time. Within the author's experience, without exception, the organizations requiring quality reports have been very successful while those having little appreciation for the information found in quality reports have had much poorer records. It should also be noted that a document of authority has the appearances of a document of authority. This is the reason that stocks, bonds, monies, and insurance policies are engraved and printed elegantly on quality paper. You cannot expect your management to think much of your personnel's reports if they lack a professional quality in style, content, or presentation. Quality information may be lost because it was presented in poor form. But one must never expect style and form to sustain misleading or poorly conceived and poorly executed research. In dealing with caking problems one is dealing with the laws of nature. Even if your management will accept poor data and poor reports, sooner or later poorly executed work is certain to have disastrous consequences for any product.

Seldom should a project be terminated without a well-written, well-documented report. Moreover, report writing is a vital part of any project and the personnel doing the work should be given the time to write a professional report. Some archival system for reports and a method of retrieving reports rapidly is also necessary. It is a waste of efforts to write a report if no one knows how to retrieve it after it has been written and filed. In years past textbooks published by John Wiley and Sons contained a flyleaf with a statement credited to Carlyle. "The true university is a collection of books." It can also be stated that the true research laboratory is a collection of well indexed reports, scientific articles and patents.

The authors of a report should have some standard format to follow and some established approval and referee mechanism. As manager of the project, it is your responsibility to know that the project has been properly conducted and that the report contains the required information to enable your management to make valid decisions as to what should be done with the reported work. Laboratory work is best understood when it is presented in a well-organized standardized report. Even the best notebooks are all but incomprehensible except to the author of the notebook.

THE WHITE POWDER SYNDROME

It is unfortunate that too many persons associated with a product can describe it only as another white powder. This can include everyone from management, to sales, to operations. Products and chemicals have personalities similar to people. A white powder may be personified as sweet, caustic, bitter, explosive, harsh, soft, light, abrasive, volatile, helpful, or harmful, just as personnel may be described. Everyone associated with a product should know its personality. To know a product is absolutely necessary. The products come alive when they are known. Once one can know how their product is better than the competition's or how their product should be improved to make it better then the competition's the product comes alive. A white powder is a product. A white powder is your product.

ENVIRONMENTAL AND HEALTH ISSUES

Today environmental and health issues must be considered in any change of a product or process. Even an old product that may be changed by nothing more than adding a trace impurity to control caking must be considered carefully before the change is made. Often additional extensive testing may be required far in excess of that required to determine whether the product is an improvement in all other regards. The expected gains from an improved product must be carefully compared with the expected cost of safety testing. It is possible that there will be no profit remaining after the cost is considered.

Even the mixing of ingredients that are known to have good safety records can be subject to extensive and expensive testing if agencies have regulations in these areas and most do. Each new item must be considered as a new case and care should be taken to make certain that you do not place your company in jeopardy by marketing a product that has not been properly cleared.

In some cases it may be better to continue to sell an inferior product that does not require new approval than it is to market a superior product that will require large sums to acquire approval. Too many new products simply cannot withstand the impact on potential earnings. In areas such as foods and pharmaceuticals, the testing has always been required to insure safety and public health and the newer regulations do not pose a problem. But the era of the backyard garage soap and paint company has come to an end. Only corporations with sufficient wealth to properly research and test a product can expect to introduce new innovations into markets and environments of the future. Most caking problems should occur in products of sufficient volume to support the work necessary to insure the public safety. If a product is not of this size, it is unlikely that research will be expended on caking problems for any reason.

TYPES OF LABORATORY TESTS

There are few analytical questions which cannot be answered satisfactorily in a well-equipped modern laboratory. Unfortunately, many laboratories do not have all of the latest and greatest equipment available but do have urgent questions to answer with the equipment they have available. It remains true that outstanding work can be performed in a tin can or a soda pop bottle, provided that the investigator has the understanding and knowledge to accurately judge his observations. This study will be directed to the laboratory with average equipment.

It is mandatory that at least two pieces of laboratory equipment are available at all times. These are a microscope and an analytical balance. The better the quality of both, the better the chances of obtaining satisfactory results will be. Some specialized equipment is required form time to time but this can usually be accessed at a local university or a commercial analytical laboratory. It is unlikely that specialized equipment should be required for long periods when solving a caking problem. It is the goal of any project directed toward caking that the problem be eliminated forever in the shortest possible time.

Pieces of laboratory equipment that can be very expensive to purchase from commercial instrument suppliers can often be assembled inexpensively from equipment found in most laboratories. The assembled equipment may or may not be as good as the commercial products but are usually satisfactory for usual needs. Instructions for the construction of several of these pieces of equipment will be included in this chapter. Certainly if your budget will allow the purchase of the commercial instruments they are usually preferred but reliable data were being collected before electronics and computers were known.

In this chapter well-known laboratory tests are reviewed. But most of the tests will probably not be found in the form needed in ASTM or similar publications. These publications are outstanding and designed to set standard specifications for military, government, and industrial applications. Tests in which molds and weights are used to induce caking are ignored. Also tests in which the cake strength is obtained by a penetrometer or similar breaking test are ignored. There is no way to equal the existing literature and to rewrite the work will accomplish nothing. Attention will be devoted to test that may be more difficult to identify or find.

SAMPLE SELECTION

Sample selection is the single most important operation in any caking study. If the sample on which a test is performed is not representative of the materials being tested, the entire exercise is a waste of time. In statistics the sample is said to represent the universe from which the sample is taken. Statistics are necessary for logical and reliable methods of obtaining samples. Most good books on quantitative analysis have a chapter or section devoted to sample selection [73]. It is recommended that one or more of these books be consulted rather than repeating a well-developed procedure. It is obvious that one sample, one sack, one car, or one batch is not sufficiently diverse to obtain reliable samples if an entire product line is having caking problems. Just how many samples should be selected depends on the system under investigation. The judgement will usually depend on time and money as much as other considerations. The money consideration should compare the direct cost of the tests to the value of the product being investigated.

There is a final point that should be considered. If the problem is of importance, it is necessary that great care be taken in selecting and instructing the personnel responsible for the selection of samples. All too often samples are collected by people who do not even understand why they are collecting them. To do a magnificent job of painting the wrong house is seldom rewarding.

It is important that the caking problem is solved, once and forever when work is completed, or it should be known why the problem cannot be solved in an economical and practical way. Caking problems have a way of going away and returning. If the system is understood, this may be prevented. It is the obligation of the investigator to explain to management what must be done to eliminate the problem, permanently. If management does not choose to follow the recommendations, it should be made clear to them *in a well-written, documented report* exactly what risks are involved in the option they have chosen. Once this has been done it then becomes a business decision and the technician should keep the records in good order in the event the problem returns. The second time around management is usually willing to commit the required resources.

STATISTICS AND EXPERIMENTAL DESIGN

The impact of Japanese goods on the American markets has caused American business to take a closer look at Japanese manufacturing methods. One of the perceived advantages of the Japanese methods was their use of statistical methods, which had been available to all of the Earth's business for fifty years or more. There is no doubt that statistics is a powerful tool and the modern computer and calculators make statistical treatments very attractive. When used properly statistics can assist in the control of quality in manufacturing processes that would otherwise be very difficult. If misused statistical treatments can become a complete waste of time. This is to say that the statistical methods cannot be expected to replace common sense, physics, thermodynamics, chemical kinetics, metallurgy, or agricultural science despite the fact that experimental design is a product of agricultural science. The statistical approach is useful in both laboratory and plant conditions but is ideally suited to plant studies.

There are a number of assumptions that lie at the heart of any statistical treatment of data. The first assumption is that the data chosen for a study represents the products under study. This involves two parts. Firstly the sample is properly collected as discussed previously and secondly, the sampled system is representative of the universe on which data is required. Another assumption is that the universe of samples can be represented by some kind of a mathematical distribution function that is more or less probability. It is assumed that the variable being measured is real and that the equipment being used for the measurements can measure the variable.

There is the story of the discovery of two chemical elements, Alabamine and Virginium. Dr. Allison discovered the elements while employing magneto-optics as a measuring tool. For a time the measurements were accepted and worldwide the two elements were added to the periodic table. As time passed the work could not be verified and the elements were removed from the periodic table. The point is simply that no amount of statistical treatment could have allowed an instrument to measure that which ultimately turned out to be an artifact. It is possible that an experimental designed experiment could have led to the ultimate results more efficiently if the instruments were better understood.

It is assumed that enough data is collected to draw some logical conclusions to the significance of variations based upon variance analyses. At times this may be a disadvantage because it is possible to collect data that is difficult to present in any format other than a statistical format. At other times it is possible to do work to fill a table of data for statistical analysis. One of the prime uses of data of the type gathered in caking studies is to convince management that moneys are being properly spent and that they should appropriate additional funds to continue the study. This can often be more important than the precise technical correctness of the data. Also in an industrial setting there is never enough data to prove a point. Everyone is attuned to the fact that ultimately an educated guess will be made to support the project or reject it. If a new plant is to be built, it is certainly not the norm that everything works exactly as planed from start-up to full production, even if that is the desired goal.

EXPERIMENTAL DESIGN AND CAKING

Experiments may be viewed from many vantage points. One scheme is to divide experiments into three general types; probes, exploratory investigations, and theoretical investigations. Probes may or may not require the collection of data that can be treated statistically. A simple "yes" or "no" may be all of the data collected in a probe. The synthesis of a new chemical compound could be classed as a probe, perhaps requiring no numerical data. Most of the science of synthetic organic chemistry usually requires a very minimum of calculations. Exploratory investigations usually are of project status and often may be an extension of a probe. They are studies in which a large number of factors and conditions may require measurements. An example is the determination of the significant factors governing the volume of a biscuit baked with a variety of ingredients and under a variety of conditions. The theoretical experiments may require precise measurements. They are usually intended to yield scientific information about some property of a system when the system has been subjected to one or more controlled variables. One example could be the measurement of the kinetics of cell division in the presence of polymeric polycarboxylates, as a function of concentration of the polycarboxylate, and the theoretical analysis and mathematical generalization of the data.

It should be noted in the science of chemistry the total variables are encompassed by composition, energy, temperature, and pressure. Structure and symmetry could be included but these are a manifestation of energy.

Caking studies may involve all three of the previous types of experiments and statistics can be applied to all three types but experimental design is most profitably employed with exploratory investigations. Top-quality production is almost impossible without the use of statistics for control and if a solid product is to be free of caking problems, it is required that the control charts be maintained and the confidence limits of the process be known and respected.

One demonstration of the statistically designed experiment is the implementation of the null hypothesis. The null hypothesis assumes that there is no difference between two or more sets of data. It is only when significant differences are found that conclusions can be drawn. This is a manifestation of the concept that it is impossible to "prove" a negative concept. This element of statistics is poorly understood by many persons in management and government. In a world gone hyper over safety, it is imperative that some way is found to teach that it is impossible to "prove" that a product is *safe*. If an infinite number of experiments are performed to prove the safety of a product, the next experiment may prove that it is unsafe. Statistical analysis recognizes this fact and sets more or less arbitrary limits. The implication as applied to caking problems is very clear. It can never be proved that the changes made to a product have made it safe. This is ridiculous but it is the world of today.

There are many good books that deal with statistics and experimental design. Many examples are presented that can be applied directly to caking problems. Again, it is much better to cite a few good references than to treat the subject in an unsatisfactory, superficial way. The original book by Fisher is required reading. Fisher was faced with a number of uncontrolled variables in studying farming yields where soil quality varied throughout the plots he studied. He devised the first experimentally designed experiments to study the variables and how they interacted [74]. Other statistics books found to be useful can be applied to a variety of caking problems with but minor modifications of the original plans [75] [76] [77] [78].

WATER ABSORPTION

Before discussing the absorption of water by solids some background on percent relative humidity needs to be discussed. It was noted that the percent relative humidity is being misused by some investigators involved in caking studies. Admittedly, percent relative humidity is a very awkward way to define the water vapor tension of air. Moreover, percent relative humidity is a poor choice for use in caking experiments. Vapor tension expresses the vapor pressure of water in the atmosphere and is independent of a stated temperature.

Water has a definite vapor pressure at all temperatures between zero degrees and the critical temperature. Most handbooks of chemistry and physics have tables of these pressures and even supercooled water is reported. Water has a solubility in air that depends on temperature and pressure; the higher the temperature the greater the solubility of water in air. The percent relative humidity is the partial pressure of water in the air divided by the vapor pressure of water at the same temperature, multiplied by one hundred to convert the fraction to a percentage. Relative humidity is then nothing more than the percentage water in the air with respect to the maximum water the air can dissolve at the chosen temperature and total pressure.

To make certain that the concept is well understood, so many measurements in caking studies depend upon understanding relative humidity, several examples will be discussed. The statement "The percent relative humidity is 60%" is not useful unless the temperature is specified. On a cold day the outside temperature is 3°C and the relative humidity is 40%. This means that the partial pressure of water in the air is 5.685mm × 0.40 or 2.274mm of mercury. At the same time the temperature in the laboratory is 26°C and the relative humidity is 10%. The partial pressure of water vapor in the laboratory is 25.209mm × 0.1 or 2.521mm of mercury. The relative humidity of the outside is four times as large as the relative humidity of the inside of the laboratory but the partial pressure of water in the laboratory air is greater at 10% RH and 26°C than it is outside at 40% RH and 3°C.

A sample of solid would be more inclined to absorb water at 10% relative humidity and 26°C than at 40% relative humidity and 3°C. Because dissolved salts lower the vapor pressure of water over their solutions, this assumes that the solubility of the salt in water is not greatly changed by the change in temperature from 3°C to 26°C.

Water is absorbed by a solid as a function of the partial pressure of water in contact with the solid under most ambient conditions. It is, therefore, much better to refer to the vapor tension of water, that is the partial pressure of water vapor in the atmosphere over the salt, than it is to refer to relative humidities unless the concept of relative humidity is well understood.

One of the first tests to be made on any sample is the absorption of water from the atmosphere. This test should begin by measuring the relative humidity of the air in the laboratory and converting it to the vapor tension of water in the air in the laboratory. Any of the more or less standard methods will be satisfactory, including a horse hair hygrometer, but it should be calibrated at regular intervals by a dew point measurement or a wet bulb thermometer. Do not rely on local weather reports for relative humidities when the measurements are made indoors!

Perhaps before leaving this subject, a discussion of relative humidity as determined by the dew point method would be helpful. All that is needed is a polished metal beaker or other thin walled metal container that will become visibly fogged when the temperature of the beaker becomes less than the dew point of air. The fog must be easily detected, as when one breathes on a cold mirror. A good thermometer, distilled water, and ice are also needed.

Fill the beaker about half full with room temperature water. First measure the temperature of air in the laboratory with the thermometer and record the reading. Next, place the thermometer in the beaker and while stirring the water slowly add ice to the beaker. When the temperature of the beaker drops low enough, water in the air will suddenly condense on the beaker, fogging the polished surface of the beaker. Note and record the temperature when the outside walls of the beaker first become fogged. The temperature just noted and recorded in the notebook is the dew point of the air. From a handbook the vapor pressure of water at the temperature just measured can be found and this is the vapor pressure of water at the dew point. Record this value. Then obtain the vapor press of water at the temperature of the surrounding air from the handbook. The relative humidity of the air in the laboratory is obtained by dividing the vapor pressure of water at the dew point by the vapor pressure of water at the laboratory temperature and multiply the quotient by one hundred to obtain the relative humidity of the air in the laboratory.

To measure the absorption of water by a sample of a product under study, select a small dish of aluminum, glass, or plastic. The dish should be about two inches in diameter and should be capable of containing at least ten grams of the sample to be tested. Determine whether the sample loses or gains weight under laboratory conditions. The time of year will influence this measurement as well as geographical location of the laboratory conducting the test. It is intended to be "quick and dirty" starting place but should give direction for the conditions for the next test. If the sample quickly absorbs water from the atmosphere, the more critical test should be done under milder conditions. On the other hand, if the sample does not gain weight or even looses weight, the next tests should be done at higher vapor tensions. By following the previous technique, it is possible determine ranges of conditions under which a product should or should not absorb water from its environment.

Most handbooks of chemistry give the details of preparing saturated solution constant humidity solutions for use in desiccators to produce small chambers of known humidity. It was noted in Chapter IV that when temperature is held constant, a two-component, three-phase system has zero degrees of freedom and the vapor pressure must be constant and fixed. It is recommended that at least three of these chambers be prepared and maintained with relative humidity values near 50%, 70%, and 85% at room temperature. The rate of change of water gain or loss should be determined as a function of relative humidity. Once these data are available, it is known what the limits of exposure of the sample to various atmospheric conditions is likely to be and how much protection the sample will require.

It is usually simpler to use systems that use saturated solutions to control relative humidity rather than preparing solutions of known concentration but that are less than saturated. There is no invariant point with an under saturated solution and the concentration of the controlled solution changes each time water is absorbed or lost. The excess of salts in the saturated solutions acts as a "buffer of state" to resist change as long as the temperature of the humidity controlling system remains constant.

PHASE CHEMISTRY

The behavior of a sample will depend in part on how it absorbs water. If the sample forms no hydrates, then all of the water that is absorbed is forming saturated solution. This will continue until the sample dissolves completely and the solution becomes dilute enough to have a vapor pressure equal to the vapor tension of the atmosphere. If the sample forms hydrates, it may absorb water until a hydrate is formed that has a vapor pressure equal to the vapor tension of the atmosphere in which it is contained. Depending upon the solubility of the hydrate, the sample may or may not continue to absorb water after all of the hydrate is formed. If the solubility of the hydrate is great, it will behave just as the soluble salt that did not form hydrates.

The data that are obtained from the previous work should be plotted as percent weight gain versus time. From these curves it should be rather easy to determine hydrate formation and similar changes from changes of slope in the plots of curves of weight as a function of time.

SOLUBILITY

If the sample is known to be a reasonably pure substance, the solubility of the sample in water should be determined if it is not already known. If the sample is a complex mixture, the solubility of the system can be very difficult to determine because it is likely to be a multi-component phase system. Fortunately, for most compounds the solubility of the pure compound has been measured and literature values are available. The higher the solubility of a compound the more likely it is to cake when it is exposed to the atmosphere.

To measure a solubility for caking studies, it is usually not necessary to know the solubility extremely accurately. A thermostatically controlled water bath, accurate to at least $+/- 0.2°C$ is needed for reliable work. If the bath is controlled at 30°C, it should perform well throughout the year without the need for external cooling water. Room temperature is seldom above 30°C but tap water, often needed for cooling a thermostated bath, can have a temperature above 25°C in some summer locations, making 25°C difficult to maintain.

A two-ounce, small necked, tared bottle with a cap that closes securely, is ideal for most solubility measurements. Fill the bottle slightly less than half full with distilled water. (If possible, a trial solution should be made first because some very soluble salts may need adjusting of the water to allow all of the sample to be added to the bottle.) Weigh the bottle and water. Record the weight of the water used. Slowly add the sample under test into the bottle being certain that the last increment has dissolved before more sample is added to the bottle. When no more solids will dissolve after a few minute's stirring, add a small excess of the solids. Tightly screw the top on the bottle, shake the bottle well, and weigh the bottle and its contents. From this weighing an estimate of the solubility can be calculated.

Hang the sample bottle in the water bath. Continue to shake the bottle at fifteen minute intervals for the next hour. If all of the solids dissolve, a small exact amount more of the solids should be added to the solution. Obviously, careful records of the changes are required. The idea is to maintain as much liquid phase as possible while in contact with some small quantity of solids. This will allow the system to reach equilibrium much more quickly than will happen if much solids are left on the bottom of the bottle. The hydration of salts can be particularly difficult if a quantity of salt must be hydrated in a saturated solution.

It is obvious that the bottle does not have to be weighed, nor does the water and the sample, but the weighings can be used as a check on the final analysis. If two solid components are being added to the bottle as in the determination of a phase diagram, then all the weighings are required.

The bottle should be shaken at regular intervals for the next twenty-four hours. A submerged magnetic stirrer placed in the constant temperature bath is often helpful to hasten equilibrium but it is not necessary. More time must be allowed for the samples to come to equilibrium if they are not stirred continuously. After twenty-four hours in the bath, the liquid phase in the bottle should be analyzed for dissolved solids. Several techniques can be used, but if it is known that the solid sample does not decompose upon heating to 110°C, it is possible to drive the water from a weighed amount of the solution followed by weighing the residue after the water has been evaporated. A one to two milliliter sample of the supernatant liquid is extracted with a hypodermic syringe and then added to a weighed 25ml short form crucible for drying, providing the solubility is not extremely low. If the solubility is very low, a larger sample may be required or a more specific and sensitive method of analysis may be needed.

It may be necessary to filter the supernatant liquid before it is placed in the crucible. If it is necessary, the liquid can be extracted from the bottle with a hypodermic syringe equipped with a hypodermic filter. Most of the time if the liquid shows no signs of suspended solids, as determined by an inspection for Tyndall effect, it may be extracted directly with a medicine dropper or similar burette and transferred to a weighed crucible.

The crucible and liquid should be weighted and transferred to a forced air oven heated to 110°C for at least one hour. After heating, the crucible should be quickly transferred to a desiccator when it is removed from the oven. The crucible may be weighed when it has reached room temperature. The solubility can be calculated from the weighings. This analysis should be repeated once each day until the sample is know to be at equilibrium. At last on the final day of testing, the sample should be analyzed by an alternate method to determine that the sample weight loss is yielding a reliable value. The methods should agree within a few tenths of a percent. Solubility data are seldom much better than this despite the great claims for perfection and dedication to duty when the data are published.

Three-Component Aqueous Phase Diagrams

There are many similarities between the determination of solubilities and determination of an aqueous phase diagram. A precisely controlled thermostated bath is a must and the temperature should be controlled to at least $+/- 0.2°C$. There are several differences that must be observed. It is simpler to avoid Schreinemakers' wet residue method of analyzing the solid phase in an equilibrium mixture when it is possible to do so [79]. But there are systems that must be analyzed by conventional methods. The reason for the wet residue analysis is to fix a point on the tie line between a solution and a solid phase. If the solid phase is either a pure compound or hydrate, enough tie lines will converge at a composition point, which defines the composition of the solid phase. This causes the analysis to be unnecessary except for confirmation when required. When a system of solid solutions forms, the wet residue method is required in most cases because the tie lines do not converge. If the wet residue method is to be avoided, it is necessary to prepare the solutions very carefully, while making each weighing precisely. Only when the system is precisely formulated can the point on a tie line be precisely located.

If the solubility of each of the solid components to be studied in a three-component aqueous phase diagram is known, the best place to start is to plot the two solubilities on the axes of a sheet of triangular graph paper. If the components form hydrates, plot the composition of the hydrates on the same axes as the anhydrous salt. In this case the percent water in the hydrate is calculated and plotted.

Prepare an estimate of the diagram by drawing a line from the solubility of the salt on one axis to the composition of the solid salt on the other axis. Draw the line from the other salt in the same way. This should form two triangles, roughly the shape of the fields in a two component diagram with an invariant point where the two lines cross. The assumption has been made that the salts will not form new hydrates, double salts or solid solutions. This may or may not be true but it is a good starting place. The assumption is made that the diagram will be similar to Figure 17 in Chapter IV.

Wash and number eight two-ounce bottles. It is usually better to number the bottles by scratching the glass with a stylus or vibrating tool. The caps should also be numbered. Dry the bottles near 100°C and when they are cool weigh each bottle and its cap. Record its weight. Set the bottles aside until they are needed.

The next step is to prepare the solutions for the trial diagram. Mark eight evenly spaced points on the graph paper about four to five percent water below the upper lines just drawn on the graph paper. These points correspond to the composition of the trial solutions. An effort is made to come close to the expected solubility of the salts while leaving a small quantity solid undissolved when equilibrium is obtained. Weigh the components into the corresponding bottles and record the weights as well as the total weight of the bottle, the cap, and the contents. Shake the bottles at room temperature until no more solids will readily dissolve. If too much solid remains in the bottle, additional water may be weighed into the bottle. If all of the solids have dissolved, more of the major component for that solution can be weighted into the bottle and the weight recorded. The major component is the component whose axis is closest to the point. The point on the triangular graph paper representing the sample must also be corrected each time a change in composition is made. No change in total composition should be made after the first sample has been withdrawn for analysis. The bottles are now ready to be placed in a thermostatic bath.

The bottles may be stirred mechanically or shaken by hand. Either way the samples should be allowed sufficient time to come to true equilibrium. No more solids should be crystallizing from solution and some solids should remain on the bottom of the bottles. If necessary readjust the compositions. When equilibrium is believed to be established, analyze the solution phase as described previously. Once the solution is near equilibrium successive samples may be withdrawn and analyzed without changing the total composition enough to influence the equilibrium values significantly. At equilibrium the precision of the replicate samples should agree at least to $+/- 0.5\%$. Since two of the three-component concentrations must be determined in order to specify the location of a point, considerable care must be exercised to obtain a precision of $+/- 0.5\%$.

When the analyses are obtained, plot the data in the following way. Locate the point on the diagram that corresponds to the analysis of that solution. Then draw a straight line between the prepared concentrations of the solution and the point determined by analysis. Repeat this procedure for three or four points. If an extension of the lines converge at a point, the lines will be in a two-phase area of a single solid phase and a solution. But if the lines fan out in an orderly fashion, the system is probably composed of one or more solid solutions. With a little practice a relatively few analyses can yield a reliable diagram but it is always wise to check to see that samples made up in the three phase areas between the dominant fields all have an equilibrium solution of the same composition as demanded by an invariant point in a diagram.

Two-Component Melt Phase Diagrams

A differential thermal analysis instrument is the most popular means of determining a melt diagram but as mentioned it does have several compromises built in to the method. If available, the commercial instruments are convenient to use but the sample size is very small on most units. Many operators are inclined to run heating curves much too rapidly, leaving no room for equilibrium to even be approached. If the sample is inclined to attack its container in any way, the same samples are usually adversely influenced by quantities of contaminants that can be ignored in larger samples.

A satisfactory differential thermal analysis instrument can be built in a short time for a nominal expenditure. The most expensive item needed to build a unit is a millivolt x-y recorder. Even a strip chart recorder can be used but the results are not as satisfactory and the strip chart recorder should be either a dual pen or multi-point type recorder. The most critical part of the unit is the thermocouple, its design and construction. Fine gauge platinum-platinum-rhodium thermocouples can give excellent results and are very simply made with an oxygen torch. (Dark blue eye glasses should be worn when welding platinum because the radiation from the metal is bright enough to damage the eyes if it is looked at for long periods.)

Little difficulty is encountered in building a differential thermal analysis units that performs well. A typical arrangement is shown in Figure 25 and can be assembled and checked for accuracy within a few hours. All of the required equipment is an x-y millivolt recorder, a good thermometer, a thermos bottle, a small pot furnace (about 2.5 inches in diameter by 2.5 inches in depth heated chamber) of the Hoskin type works well, and a variable transformer large enough to handle the current supplied to the furnace. The normal 120-volt variable transformer is usually sufficient for use in the United States.

Some special consideration must be given to the sample container and the temperature reference container. Perhaps the most simple approach is to use a thick-walled Alundum crucible to hold the sample crucible.

The sample crucible may be made of a ceramic or platinum and should fit snugly into the Alundum crucible. The Alundum crucible being chosen to fit into the furnace chamber loosely. The Alundum crucible itself will be chosen as the temperature reference to which

Solubility / 155

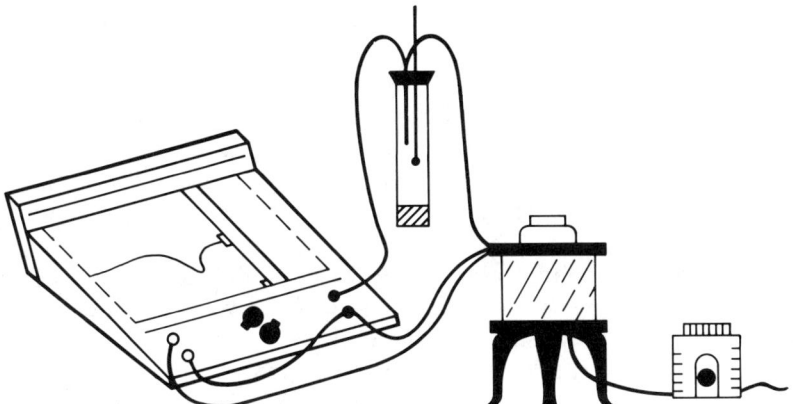

FIGURE 25. *A laboratory constructed differential thermal analysis unit with x-y recorder, room temperature standard, furnace and transformer.*

the sample is compared. This will be done by drilling a small thermocouple well into the walls of the crucible from the top side. A small drill bit about 1/16 inch or smaller will be satisfactory and the hole is made with either a diamond or carborundum abrasive to grind the well for the thermocouple. The arrangement is shown in Figure 26.

A second approach is to cut a 2 inch long thermal block from a

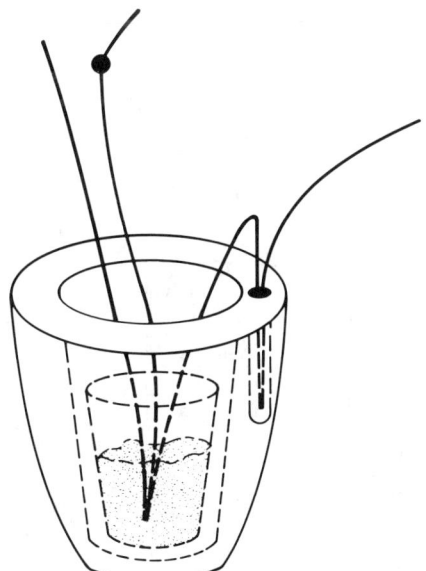

FIGURE 26. *Cut away view of sample holder with thermocouple well in an alundum crucible used as a reference temperature.*

1.75 inch rod of circular bar stock nickel metal or high nickel stainless steel. In this case the block is drilled with two ¾ inch holes 1.5 inch deep as shown in Figure 27. The crucibles containing the samples and reference materials are placed in the holes of the thermal block. The advantage of this approach is that the base line can be controlled more easily because the heat is more evenly distributed but almost equal results can be obtained in either method. The reference material was either aluminum oxide or electronic grade magnesium oxide for some samples if magnesium oxide matched the heat capacity and heat transfer of the sample more closely. The reference material cannot have a phase transition in the temperature range of interest.

The temperature range of the thermal analysis unit is from about −30°C to about 1000°C and it should be calibrated at several temperatures by measuring the melting temperatures of several standards. The transition temperature of sodium sulfate decahydrate has been mentioned several times and is an excellent 32°C standard while the 398°C melting of potassium dichromate is very sharp and vivid. Other standards can be chosen as needed. In all determinations the temperature of the thermometer in the thermos bottle should be read and recorded. The room temperature thermocouple is designed to buck out the room temperature reading and temperature-voltage ta-

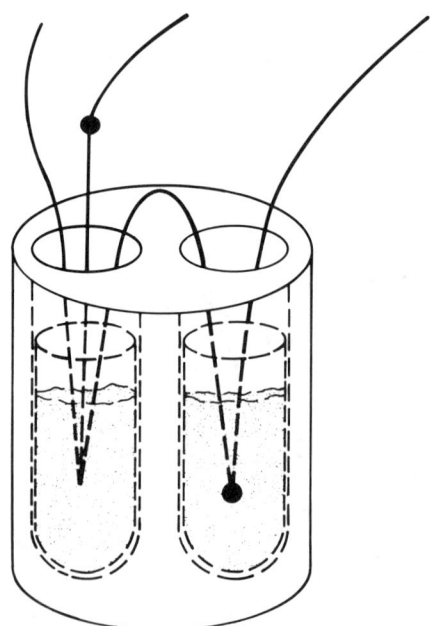

FIGURE 27. *Cut away view of sample and reference holder made from a drilled stainless steel block. Reference is in the right side of the block.*

bles used to convert voltages to temperature can be read as zero degree references without added corrections. In most cases the equipment can be expected to yield results about $+/- 1°C$ but this is better than most phase diagrams can be assembled, despite the literature claims. As more information may be required two books are suggested. *Handbook of Differential Thermal Analysis* is excellent and gives many details of thermocouple placement and temperature readings. *Temperature, Its Measurement and Control* is also outstanding for many uses [80].

In preparing thermocouple junctions *do not* twist the wires to join them and do not use mercury connections to weld the wires. This will put strain in the wires to twist them and mercury and platinum quickly form an alloy with characteristics unlike pure platinum. Select one wire of platinum and one wire of platinum-rhodium. Either ten or thirteen percent rhodium thermocouple wires may be used but the preference is usually ten percent rhodium wires because they are more easily accessible in most laboratories. In order to limit the heat capacity of the thermocouple wires near 18 gauge work well but slightly larger or smaller wires can be used. Light the oxygen torch, which may be fueled with either butane, propane, natural gas, or hydrogen, however, butane is the preferred fuel. Straighten the wires if necessary. Place the wires side by side holding the tips of the wires together. After making the two and three wire junctions, small diameter ceramic beads or tubes should be put over the wires far enough to insulate the wires as they exit the furnace. Out in the room any type of insulation, for example, fine diameter rubber tubing or the like may be used to prevent the wires from touching. The final thermocouple is the room temperature bucking potential thermocouple. It is prepared as the other junctions and the wire should be insulated up to the point where it is connected to the recorder.

(Remember to use the dark blue eye glasses.) Heat the tips of the wires until a small, white-hot, bead is formed that will weld the wires as it cools. Using this technique prepare the thermocouple shown in Figure 28. This thermocouple harness has several advantages. Firstly, it has a room temperature couple that can be used to automatically adjust for a room temperature reference.

As shown the reference temperature must be subtracted from the value read on the recorder. It will be seen that if the reference thermocouple is made in the other leg of the measuring thermocouple, it can act as an automatic bucking thermocouple and the temperature can be read directly from the tables. But the next advantage will be forfeited. Secondly, since all wires leading from the thermocouple are of the same composition, in the case shown platinum,

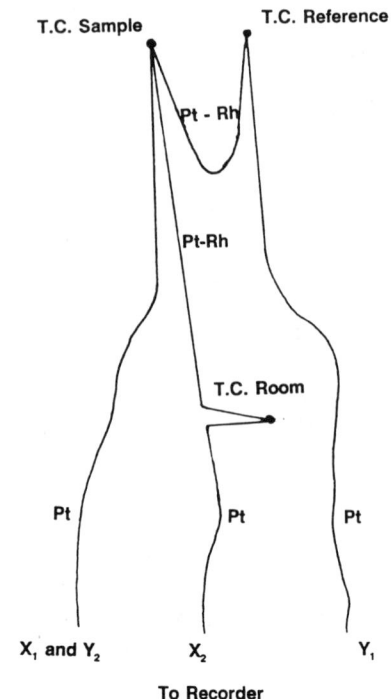

FIGURE 28. *Thermocouple design used with the differential thermal analysis unit shown in Figure 25.*

contact potentials with the recorder metal connections all cancel. Lastly, ice point reference temperature is eliminated because the room temperature thermocouple will serve this purpose very well. The fine diameter wire is inexpensive enough to lead the wires from the furnace to the recorder without the need for compensating wire but since all lead wires are of the same metal, copper lead wires can be used provided all junctions are maintained at the same temperature. (Because many errors have been made as a result of this error, it will be repeated. Platinum-rhodium wires are commonly 10% or 13% rhodium and are used as one wire while pure platinum is used as the other wire. Either type rhodium-platinum wire can be used but it is required that the type is known. If necessary, the type can be determined by calibration of the completed thermocouple. Also the platinum-rhodium wire is much stiffer than the pure platinum wire.)

Note the action of the thermocouples. The temperature of the sample is read across the two wires leading from the sample including the wire containing the room temperature thermocouple and is recorded through the horizonal axis connectors of the recorder. The differential temperature between the sample and the reference is read across the leads that do not contain the room temperature thermocouple junction and are plotted vertically on the x-y graph. A jumper wire should connect the x and y inputs of the recorder to complete the circuit of the differential thermocouple. The jumper is connected to the recorder x input that connects the wire form the sample thermocouple that does not contain the room temperature thermocouple. If on the recorder the polarity is reversed, simply exchange the leads to the recorder. Most investigators prefer that endothermic reactions are recorded as a dip in the temperature curve.

Each time the unit is used, it is necessary to coordinate the recorder and the temperature on the graph. This is simply done. Place the sample thermocouple, the room temperature thermocouple, and the thermometer together in an isolated box, cloth, or other protective covering that will allow them to come to the same temperature and after reading the potential from the thermocouple chart adjust the recorder to this voltage with the x control on the recorder. Of course the proper voltage range must be chosen to cover the expected range of temperature in the determination and the polarity of the reference temperature must be observed because most tables of temperature versus potential assume a zero degrees reference temperature.

Because the differential thermal analysis units that are built in the laboratory may at times require an ice point as a thermocouple standard, a discussion of the ice point is in order. The ice point (0.000 +/− 0.001°C) is often misused. An ice cube floating in a bucket of water does not establish zero degrees Celsius even if the bucket is a Dewar flask. To obtain zero degrees, it is necessary to use pure ice that has been converted to a snow and then saturated with clean air. Transparent commercial ice is an excellent source of very pure ice. The water obtained from the transparent portions of commercial ice approaches conductivity water in purity and is far superior to most distilled or deionized water. Saturating the ice-water mixture with air lowers its temperature by 0.0024°C. It is an *equilibrium* one component, two-phase system with pressure held constant at atmospheric pressure in an open vessel. In very precise work, changes in atmospheric pressures require corrections. Thermal work can be in error by degrees if an improperly prepared ice point is used blindly as a standard. Commercial electronic standards can be used and are superior to poorly prepared ice point standards.

The triple point of water is a much better temperature standard for laboratory use. The difference is that the triple point is a one component three-phase system contained in an evacuated container (no air included) where the pressure in the system is the vapor pressure of water at the temperature of the triple point. If all three phases remain in contact, then both temperature and pressure must remain constant at equilibrium (0.0099 +/− 0.0001°C). Caking studies will not usually require this accuracy but it is better to have a fixed laboratory standards overly defined than to rely on a standard that is varying erratically. This is particularly true when preparing phase diagrams by differential thermal analyses.

WEIGHT LOSS ON IGNITION

If a thermobalance is available, it is an ideal tool for the study of weight loss on ignition. Several factors can contribute to successful use of the thermobalance in this analysis. The larger the sample under study, the better the analysis will be. The rate of heating of the sample is also very important and the slower the sample is heated, the better the results are likely to be. A two-hour heating period is not unreasonable and in some cases if several hydrates are to be differentiated more than two hours may be required [81]. The upper temperature to which the sample should be heated will depend upon the properties of the compounds. In most cases the samples should be heated to at least 250°C in order for most hydrates to decompose. Other samples should be heated to higher temperature if the sample can be heated without some undesired charring or similar reactions taking place.

If a thermobalance is not available, the sample can be weighed into a preheated to constant weight, 25ml short form crucible and then heated to 100°C for about thirty minutes. Transfer the crucible to a desiccator. After the sample has cooled, determine its weight and place it back in a furnace set at 125°C and heat for thirty minutes more and redetermine the weight loss. Raise the temperature as many times as is required to determine whether the sample is a hydrate and how much water is contained in the hydrate and how much free water is contained in the sample.

The previous treatment is good for unbound water and most hydrate water, but if water of constitution is involved, then just heating the sample is not enough. Water of constitution is the hydrogen and hydroxyl water which may be contained in an anion. As an example, NaH_2PO_4 contains one water of constitution. This water can be very difficult to remove, quantitatively, even at temperatures as high as 650°C. It is usually much better to use a basic fusion technique heating the sample in an excess of prefired ZnO [82].

In this case a few grams of ZnO are heated in 25ml short form crucible to a temperature near 600°C for at least one hour. The hot crucible is transferred from the furnace to a desiccator. When the crucible cools to room temperature, the crucible and ZnO are weighted. About 0.5g of the sample is carefully stirred in to and down under the ZnO with aid of a small wire. Any solids hanging to the wire should be brushed back into the crucible with a small camel's hair brush. Refire the crucible to a temperature high enough to cause the water of constitution to react with the ZnO but not high enough to char, oxidize, vaporize, or otherwise destroy the sample. Most inorganic solids can be heated to 550°C without creating problems. Cool the crucible as before and weigh it again. The weight loss should be all other water plus the water of constitution. The procedure works best for nonvolatile inorganic compound but it may be used with some organic salts provided the integrity of the salt is not violated.

X-RAY ANALYSES

A standard x-ray powder pattern is usually sufficient to obtain much useful information about a sample. A hot stage x-ray system is even more useful for caking studies. The reader is referred to the book edited by Kealble for instructions for all types of analyses [83]. For the identification of crystalline compounds, the x-ray technique reigns supreme. More useful information can be obtained quickly by a powder pattern than by any other single method. The limits are the fact that several percent of a substance must be present to obtain a reliable quantitative estimate of its presence and the approach is almost useless for the estimation of amorphous substances. Sometimes high base lines can be an indication of amorphous solids. The microscope can also be very useful in estimating the quantity of amorphous material in a sample.

MICROSCOPIC STUDIES

Perhaps the most useful tool for studies dealing with caking is the microscope. It is even more useful if it is equipped with three inexpensive devices. The polarizer and analyzer add much to the understanding of what is happening between particles and the structure of the sample is revealed in greater detail. An inexpensive hot stage adds to the study of phase transitions and a good hot stage can be constructed from simple parts if one chooses to do so. The third helpful tool is a camera for the microscope. Either a 35 mm camera or a larger Polaroid camera can be used. The Polaroid camera is usually desirable in this case but very good results can be obtained with an Olympus 35mm single lenses reflex camera body while using the automatic exposure system of the camera. Outstanding results were obtained even with bad lighting conditions. This approach makes high quality micrographs easily affordable.

A new book, devoted to the use of the microscope to obtain photographs, has been published John Gustav Delly of Eastman Kodak. The book is well written and describes in detail what is needed to obtain professional results. Most users will not push the art to the limit but it is well to know what can be done [84].

If an x-ray instrument is unavailable, it is possible to definitely identify crystalline substances with a polarizing microscope and refractive index immersion oils [85]. The microscope method to identify crystals is time consuming and tedious. As mentioned in the following material the immersion oils are useful in other ways for obtaining photographs or oil immersion objects. Not all oils are suitable for oil immersion lenses and care should be made to be certain that an oil is recommended before it is used.

In order to obtain the most from the use of the microscope, several techniques should be mastered. To obtain good pictures of crystal it is usually desirable to use an immersion oil and a cover glass. Oils with a wide range of refractive indexes can be purchased from a scientific supply house. Most of the time one desires that the refractive indices of the crystals and of the medium be as different as is practical. This gives the maximum relief and allows the crystals to be vividly portrayed. If the refractive index of the oil and the crystal are nearly the same, the crystal becomes almost impossible to be seen with ordinary light. The cover glass not only helps to protect the microscope objective, it also keeps the sample in a thin plane, which improves the focus of any field. For completeness, it should be mentioned that immersion oils are not usually used with hot stages. Long focal length objectives can be purchased for most microscopes. This allows a much longer distance between the hot stage and the objective.

Refractive Index

The index oils can be most useful when used with Becke line method. The Becke line method is simple to use and with a little practice refractive indexes can be determined precisely [86]. The use of the refractive index of with amorphous solids is particularly useful since the x-ray is of little value. Additionally, the refractive index of a glass is usually easily measured because the solids are isotropic and the refractive index will depend upon composition only. This means it can easily be determined whether or not a glassy phase has changed in composition as a number of similar samples are prepared. If the composition of a glass is known, it is very easy to estimate the refractive index from the molar refraction of the metal oxides and other constituents of the glass because the molar refraction of most constituents found in glass can be found in standard handbooks of chemistry and physics. These estimates are very close to the measured values, usually agreeing to at least three decimal places.

It is very difficult to obtain good photographs of crystals on a microscope stage while using reflected light. Most of the time it is much more rewarding to use transmitted light. Unless very much care and experience is employed, the specular nature of most crystal surfaces will cause the photographs to be of poor quality as though the camera lens is not well focused. If the crystals can be photographed under a liquid, better pictures can often be obtained.

Television Equipment

Modern television equipment has become so inexpensive and so splendid in quality that some outstanding results can be obtained for a nominal expense. A black and white television camera is easily attached to an optical microscope. The signal is passed through a monitor and then to a video recorder. If it is desired, the video signal can be digitized and printed in color with the aide of a home computer and color printer. In the previous case, the equipment was all home owned and had been used in other hobbies but there is almost no limit to what can be done. When the hot stage and the camera are combined, it is possible to observe and record phase transitions directly and most cameras are equipped to allow one to superimpose a voice commentary of what is happening at the moment it occurs.

ELECTRICAL PROPERTIES OF CRYSTALS

Many books have been written about the electrical properties of crystals [87]. The subject can become very involved, and it will be treated in a manner sufficient for most uses but without doing justice to the science. For the most part an attempt will be made to determine whether or not a crystals of interested exhibits piezo-, pyro-, or ferroelectric behavior and whether or not the behavior is sufficiently strong to account for a caking problem.

The hierarchy of electrical crystals is piezoelectric crystals are the most plentiful and twenty of the thirty-two possible crystal point groups exhibit piezoelectrical behavior to some extent. Pyroelectric crystals follow with ten of the twenty piezoelectric exhibiting pyroelectric behavior. Lastly, the ferroelectric crystals depend upon the formation of ferroelectric domains similar to ferromagnetic domains for their behavior. Only crystals, which have no center of symmetry, can exhibit the piezoelectrical behavior. It is also true that only crystals, which lack a center of symmetry, can exhibit pyroelectrical or ferroelectrical behavior.

Crystals which exhibit birefringent behavior in the polarized light field of a microscope have two or more refractive indexes. The first simple test to make is to determine whether or not the crystals of interest are birefringent when viewed with a polarizing microscope. If they are birefringent, they will light up and then blink dark and then light up again when they are rotated on the stage of a polarizing (petrographic) microscope. Fourteen of the twenty crystal types exhibiting piezoelectric behavior are birefringent. All crystals exhibiting optical activity (birefringence) are piezoelectric, but not all piezoelectric crystals exhibit birefringent behavior.

The subject of crystal symmetries can become very involved and is mostly reserved for dedicated crystallographers but the use of this fascinating and important area of science is not difficult. Without some knowledge of the subject attempts to solve caking problems are likely to be futile. (Unfortunately, the author cannot be considered to be more than admirer of the work of the crystallographers, but he has used their help, almost daily, for many years.)

The test for a birefringent crystal (also called double refraction) is a very quick and easy test to run. Crush a small quantity of the crystals for observation, if crushing is necessary, and place them on a microscope slide. Place a cover glass over the crystals and add a drop of immersion oil at the edge of the cover glass in a manner such that the oil will be drawn under the cover glass by capillary attraction. Place the slide on the microscope. Employ a low powered objective to focus on the crystals. Set the polarizer at the bottom of the light path to the zero position (parallel to the base of the microscope). Then while looking through the eye piece, turn the analyzer (the top polarizer) to zero if it is adjustable. Some analyzers are adjusted to zero at the factory. Rotate the polarizer 90° to give a dark field. Birefringent crystals will appear bright and perhaps colored against the dark field. It should be mentioned that crystals can be too small to exhibit birefringent behavior even though they are birefringent. If the stage of the of the microscope can be rotated, the birefringent crystals will blink on and off as they are rotated on the stage. If the stage is fixed, turn the direction of the slide with respect to the stage and note the crystal behavior. If a crystal is birefringent, it will be piezoelectric to some degree. There is also a lesser chance that the crystals are pyroelectric and even smaller chance that the crystals are ferroelectric. It is possible to identify a crystalline substance with a microscope but rather than pursue this line of testing in depth it is probably better to consider the test for the different types of electric behavior if a sample is birefringent. If a sample is not birefringent, it must be tested for piezoelectric behavior by the more direct tests. The sample may be either amorphous or is an inactive crystalline form.

Piezoelectric Test

Although there is nothing difficult in obtaining a test for piezoelectric activity in a crystal, most laboratories interested in work-a-day caking problems are not too likely to have the required equipment for the most sensitive tests but the equipment can be inexpensively built. An oscilloscope should be substituted for the audio equipment suggested in the reference [56]. Fortunately, only those crystals exhibiting a very strong piezoelectric are of interest as contributors to caking. It is usually possible to obtain a signal from these crystals by merely placing them under strain between the leads of a high sensitivity galvanometer or the leads to a high resistance volt meter. The tests can be as simple or as complex as required.

The test described below are very simple and will work for strong to medium signals. If more sophisticated test are required, several references are cited. Two simple experiments should be performed to demonstrate this technique. Select a well-formed crystal of monoammonium orthophosphate. A high impedance volt meter should be attached to two leads with small metal clips. Attach a lead to each end of the crystal by simply clamping it with the clips. Set the voltmeter on one of the low DC ranges (one volt or less full scale). Insulate the connection or put on a pair of thin rubber gloves. Now twist the crystal by placing a torque on the two clips. Note that an easily detected potentials can be seen on the volt meter and that the sign of the potential is reversed as the direction of the torque is reversed. In the 1930's and 1940's many inexpensive phonograph record players were used in which the needle cartridge contained a small crystal of orthophosphate. The crystal was mounted in such a manner that as the needle moved from side to side in the record grooves an electrical potential was generated at the ends of the crystal. As the needle changed direction, the sign of the charge on the ends of the crystal changed. Electrodes were connected to the ends of the crystal and the charge was conducted by wires to an audio amplifier, thus, converting the motion into sound.

It is slightly more difficult to demonstrate that a charge is also generated on a crystal of monoammonium orthophosphate when the crystal is pressed between two electrodes than it is to demonstrate the charge when the crystal is subjected to torque. If too much force is applied, the crystal may shatter before the charge is registered. There are much more reliable and sophisticated methods of demonstrating piezoelectric behavior. It should be mentioned in passing that not all seven piezoelectric crystals respond as monoammonium orthophosphate and more sensitive tests must be used. But at the same time as the influence is decreased the contribution to caking also decreases.

The way crystals are stressed must be considered. Piezoelectric behavior results from a displacement of charge within the crystal. The force applied must be in a direction to cause the charges to become displaced. An excellent explanation of the mechanistic origin of the piezoelectric effect is given by Van Der Ziel [88].

Pyroelectric Test

Kundt's pyroelectric activity tests are very simple and is most vivid [56]. There are two very interesting tests for pyroelectric behavior. First, heat a few of the crystals to be tested on a microscope slide or similar glass plate. It is best to heat the crystals fifty or more degrees if it is practical to do so. Next, using a blow pipe covered at the end with muslin cloth, blow some red lead oxide through the cloth and on to the cooling crystals. Then repeat the powdering except this time blow powdered sulfur through muslin and on to the cooling crystals. The result is active crystals are red on one end and yellow on the other. The red lead oxide is charged positively as it passes through the muslin cloth while the sulfur is charged negatively. Of course the test confirms pyroelectric behavior only if the opposite ends of the crystals are red or yellow.

The second test for pyroelectricity is extremely informative, demonstrating crystal properties that are all but unexpected. In this case a few crystals are prepared as before by heating them on a glass plate. Again, there is no exact temperature range through which the crystals should be heated provided the integrity of the crystal is respected. The cooling crystals are then placed under a bell jar or a two liter beaker, which has had the lip cut away to allow the beaker to rest flat on surface when inverted. In the top end of the bell jar a one-inch strip of one-eighth inch magnesium ribbon is suspended by a tape that has been secured to the top of the bell jar. The bell jar is now placed over the warm crystals and is raised just enough to allow one to ignite the magnesium strip. Then while the strip is burning, the bell jar is placed back over the crystals allowing the newly formed, highly charged, magnesium oxide, dust to settle over the crystals. Because the pyroelectric crystals are highly charged, they have a well-formed electric fields that surrounds the crystal in much the way that a magnetic field surrounds a magnet. The highly charged magnesium oxide dust settles into the field, outlining the field as iron powder would outline a magnetic field. See Figure 29. Well-grown crystals of

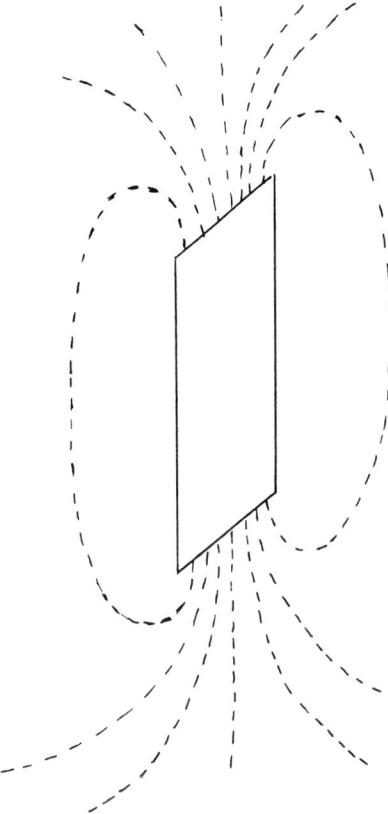

FIGURE 29. *Pyroelectric crystal exhibiting an electric field as detected with freshly formed magnesium oxide.*

tartaric acid, sodium tartrate, ammonium phosphate, or ammonium nitrate exhibit this property and make an interesting demonstration. Care should be taken not to burn too much magnesium ribbon in the bell jar. An excess of magnesium oxide can over coat the crystal and field, obscuring a positive result. A positive result is one in which the electric field can be detected. When properly performed, the results are obvious. One other word of precaution. The bell jar or beaker should fit tightly enough on the surface on which it rest to prevent outside currents of air from entering the chamber. Excessive turbulence can destroy the test.

Ferroelectric Test

The ferroelectric tests are not as straightforward as the piezoelectric test and the pyroelectric test. In order to exhibit ferroelectric behavior, a crystal must exhibit spontaneous polarization when cooled through a transition temperature referred to as the Curie point or temperature just as in ferromagnetic systems. At the transition temperature there is a very large rise in the dielectric constant of the crystals as the crystals are cooled. If the crystals are cooled through the transition temperature while they are subjected to an electric field, several interesting events happen to increase the polarization of the crystals. The first is the ferroelectric domains in the crystals are oriented more nearly in one direction and there are fewer but larger domains. One can observe the growth of domains with a microscope if the crystals are subjected to a direct current field that can be varied in potential.

One simple test is to subject a crystal that has been neither heated or cooled to the pyroelectric test. If the test is positive as a result of spontaneous polarization, there is good evidence that the crystal is ferroelectric. The crystal can then be poled by subjecting it to a DC field followed by a second pyroelectric test. If the test is stronger than before, the crystal was poled more evidence of ferroelectric behavior is demonstrated.

Yet another slightly more complex experiment is to test the polarization hysteresis of crystals subjected to alternating fields as demonstrated by Sawyer and Tower in 1929 [89]. In this experiment an oscilloscope and a simple circuit using a 60 cycle electrical source is all that is required. The crystal is placed in the circuit as the dielectric in a capacitor. The hysteresis curve is easily observed when the crystal is ferroelectric. With the proper calibration, dielectric constant can also be calculated.

SUMMARY

Laboratory tests that could be applied to caking problems are almost endless. The ones chosen for this work are some that are not usually encountered in the form presented here. The standard test are available from ASTM and other publications and should be applied as required.

Throughout this work, quality and customers have been emphasized. The customer is the final test of any caking problem. If the customer views a product as a quality product, it is a quality product. But if the customer views a product as defective when compared a competitor's product, the productive is defective irrespective of any laboratory tests which may be concocted. Like most items there is a subjective component as well as an objective component in all tests for quality.

REFERENCES

73. Bicking, C.A., *Principles and Methods of Sampling, Treatise on Analytical Chemistry 1*, Part 1 Ed., Kolthoff, I.M. and Elving, P.J., John Wiley and Sons, New York, p. 299 (1978); Mandel, J., *Statistical Methods in Analytical Chemistry, Treatise on Analytical Chemistry 3*, Part 3, Ed. Kolthoff, I.M., Elving, P.J., and Stross, F.H., John Wiley and Sons, New York, p. 79 (1970).
74. Fisher, R.A., *The Design of Experiments*, Oliver and Bond, London and Edinburgh (1947).
75. Davies, O.L., *The Design and Analysis of Industrial Experiments*, Oliver and Bond, London and Edinburgh (1954); *Statistical Methods in Research and Production*, Oliver and Bond, London and Edinburgh (1949).
76. Bauer, E.L., *A Statistical Manual for Chemists*, Academic Press, New York (1971).
77. Gore, W.L., *Statistical Methods for Chemical Experimentation*, Interscience Publishers, Inc., New York (1952).
78. Freeman, H.A., *Industrial Statistics*, John Wiley and Sons, New York (1942).
79. Schreinemakers, F.A.H., *Z. phys. Chem.* 11, 76 (1893).
80. Roeser, W.F., *Temperature Its Measurement and Control in Science and Industry*, Reinhold Publishing Corporation, p. 80 (1941); *Temperature Its Measurement and Control in Science and Industry 4*, Parts 2 and 3, ed. Rubin, L.G., Anderson, A.C., Janssen, J.E., and Cutkosky, *Thermocouples*, Instrument Society Of America, Pittsburgh (1972); Somthers, W.J. and Chaing, Yao, *Handbook of Differential Thermal Analysis*, Chemical Publishing Company, Inc., New York (1966).
81. Griffith, E.J., *Anal. Chem.* 29, 198 (1957).
82. Griffith, E.J. and Callis, C.F., *J. Am. Chem. Soc.* 81, 833 (1959).
83. Kealble, E.F., *Handbook of X-Rays*, McGraw-Hill Book Company, New York (1967).
84. Delly, J.G., *Photography Through the Microscope*, Eastman Kodak Company, Rochester, N.Y. (1988).
85. Buerger, M.J., *Elementary Crystallography*, p. 183, John Wiley and Sons, New York (1956).
86. Wahlstrom, E.E., *Optical Crystallography*, p. 56, John Wiley and Sons, New York (1960).
87. Taylor, G.W., Gagnepain, J.J., Meeker, T.R., Nakamura, T., and Shuvalov, L.A., *Piezoelectricity*, Gordon and Breach Science Publishers, New York, N.Y. (1985).
88. Van Der Ziel, A., *Solid State Physical Electronics*, p. 498, Prentice-Hall, Inc., New York, N.Y. (1957).
89. Sawyer, C.B. and Tower, C.H., *Physical Rev.* 35, 269 (1930).

CHAPTER SEVEN

Flow Schemes to Classify Caked Solids

TYPES OF SCHEMES

In this chapter there is an attempt to formulate a flow scheme that is similar to the old qualitative analysis flow scheme or modern computer flow charts. The objective will be to separate caked solids into groups and subgroups and to ultimately decide which type of interactions hold the particles together in the caked solid.

In any work directed to the identification of the cause of caking in industrial solids, it is at first mandatory to establish that a caking problem does truly exist. This can best be done by contacting those claiming a caking problem exists and request samples as well as a description of the conditions under which the sample is believed to cake. Nothing takes the place of first hand examination and the investigator is encouraged to visit the site of the problem if it is at all practical to do so.

Once it has been established that a caking problem does probably exist, the following procedure can be followed if the product is worthy of the work necessary to solve the problem. The next two items, which must be faced, are sampling and transporting the samples. If the samples are not representative of the production during caking, very little useful information can be gained. Also, if the samples are improperly packaged for transporting and gross changes occur along the way, it will be difficult to establish much information about the caking problem. These are only common sense instructions but all too often samples are transported in a paper bag and expensive, inconclusive data are collected on the wrong problem.

SCHEME OF ATTACK

In years past most universities taught a course in the qualitative analysis of inorganic cations. The analytical procedure was a rigid set of rules that when followed exactly would usually separate cations into groups, subgroups, and ultimately the individual ions. The exercise taught much about the solution chemistry of inorganic ions.

Most chemists are familiar with qualitative analysis schemes to separate and identify a number of ions or molecules in an unknown sample. Certainly knowledge is not great enough and the possibilities are too great to allow us to set up a caking scheme with anything near the reliability of a qualitative analysis scheme, but there are a number of test that can be performed in a logical order to at least identify the nature of the problem, its severity, and the type of program necessary to determine whether or not the problem can be solved.

In the following section, seven schemes have been presented. It is not the objective of the schemes to solve caking problems per se but to classify the type of caking responsible for the problem. When a sample has been tested with the schemes, it should be relatively easy to classify to which of the four types of caking the sample belongs. Once it has been established what is the nature of the problem, the problem can be attacked with the confidence that the proper variables are being studied. This does not mean that all caking problems can be solved under all conditions but it does allow one to state that logical approaches to the problem have been considered a realistic estimate of the resources required to attack the can be made.

172 / Flow Schemes to Classify Caked Solids

SCHEME NO. 1

Selection of Samples

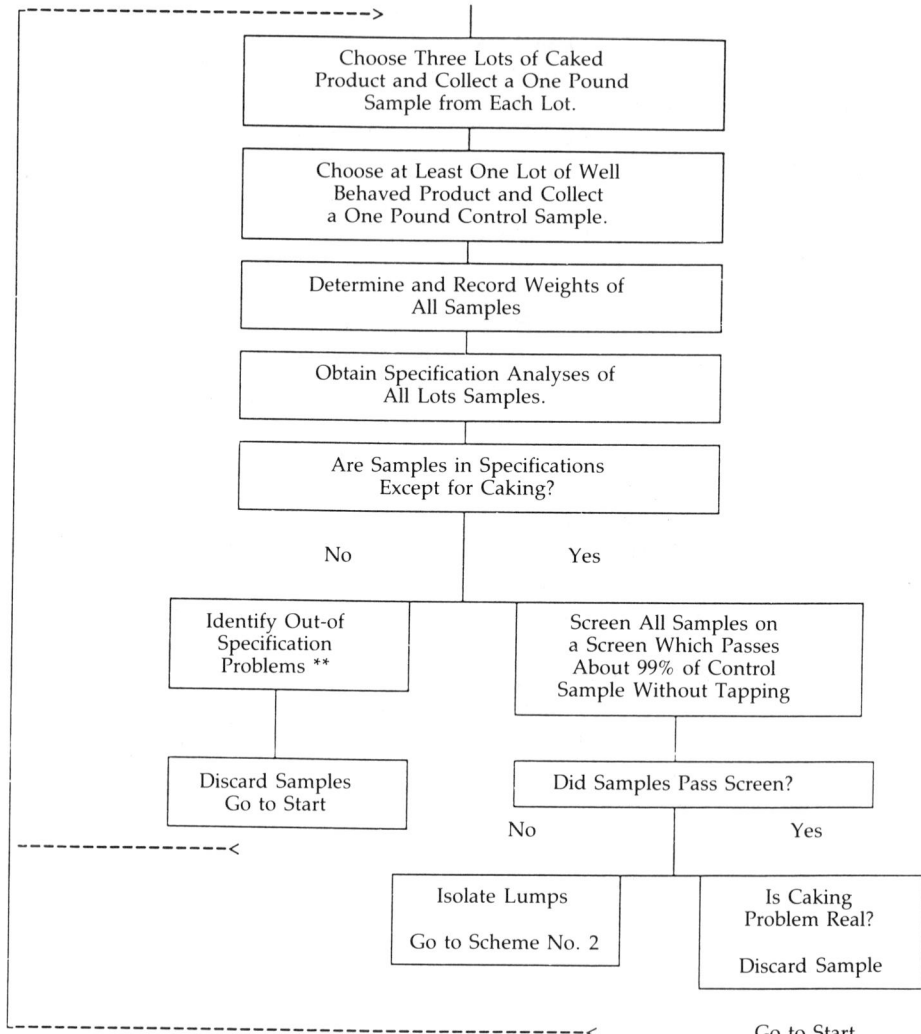

**It is assumed that caking is occurring in products that are in specification. If this is not the case, then the problem is a quality problem and not merely a caking problem.

Scheme of Attack

SCHEME NO. 2

Characterization of Lumps from the Screen in Scheme No. 1

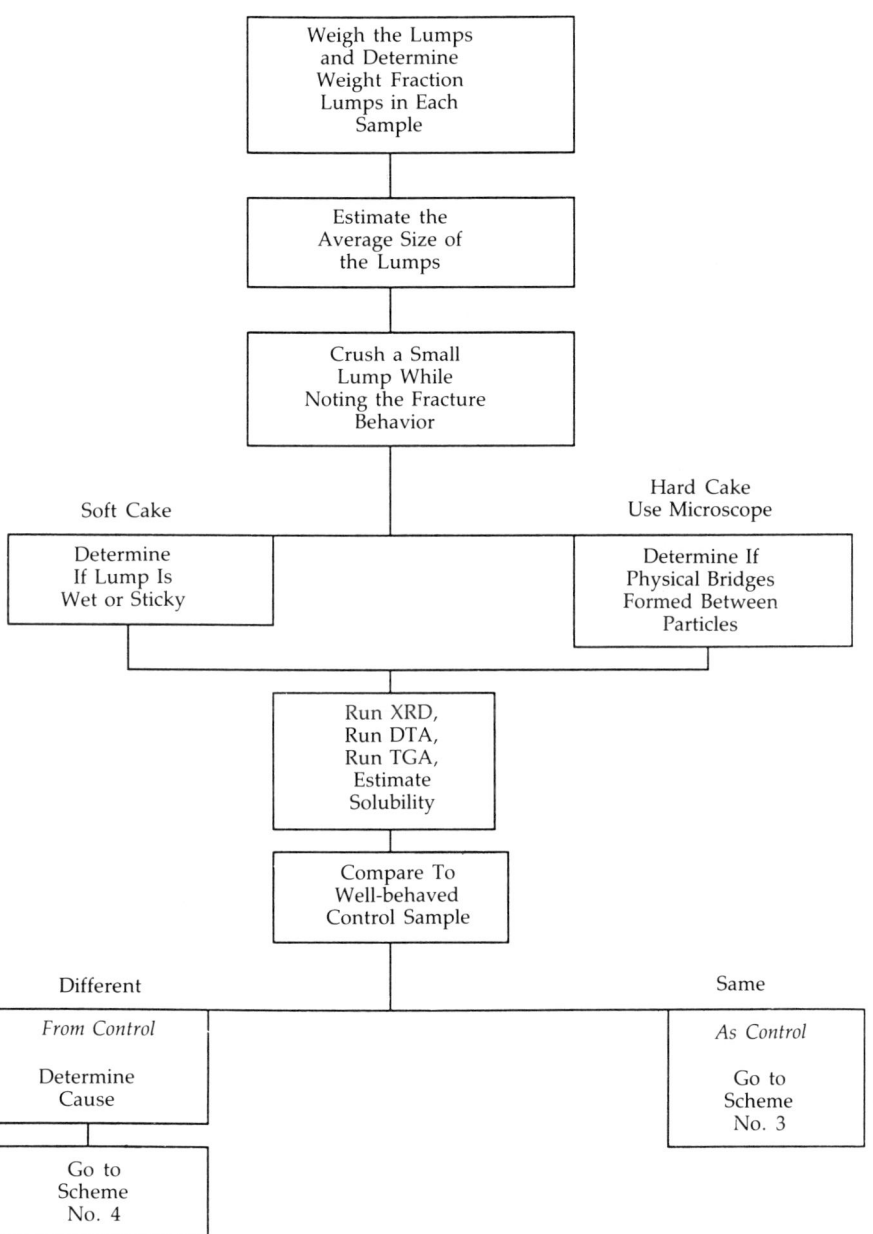

174 / Flow Schemes to Classify Caked Solids

SCHEME NO. 3

Mechanical Caking

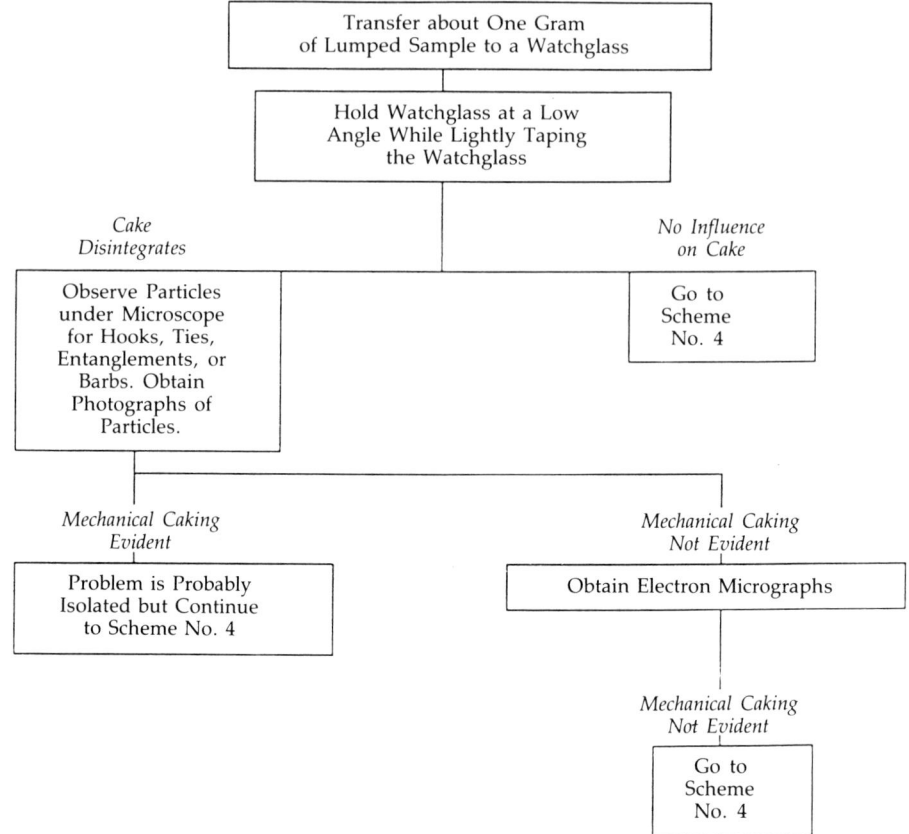

Scheme of Attack / 175

SCHEME NO. 4

Plastic Flow Caking

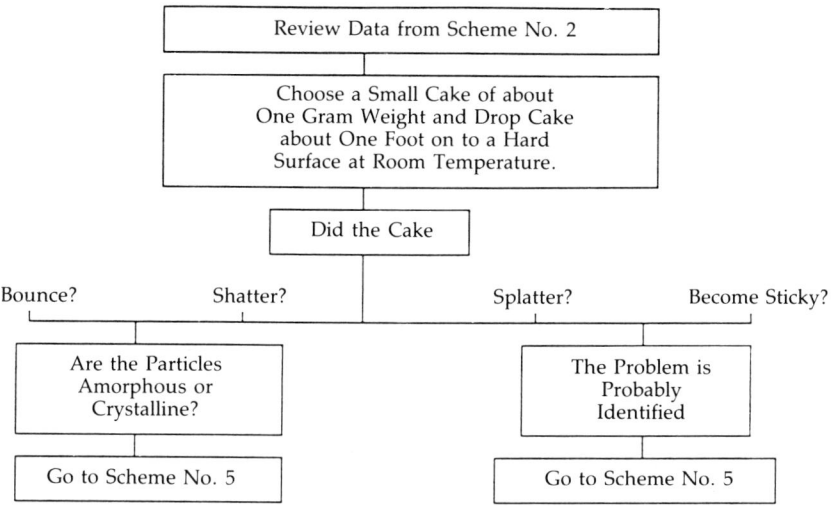

SCHEME NO. 5

Electrical Caking

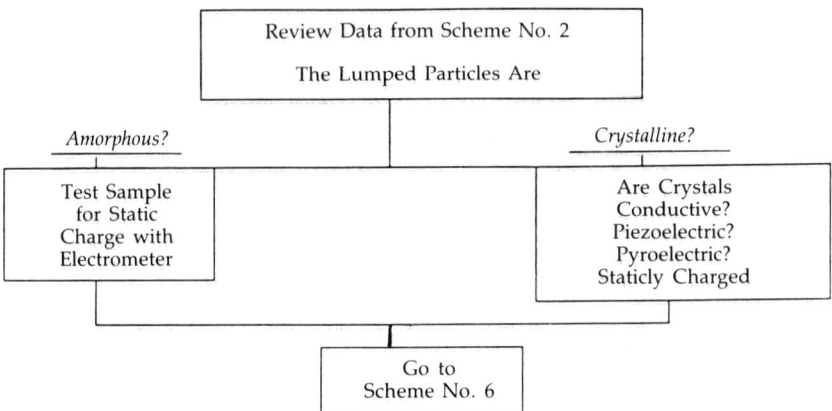

These tests are described in the text under the section Electrical Tests, *Chapter VI. The tests are too involved to include in the schemes.*

SCHEME NO. 6

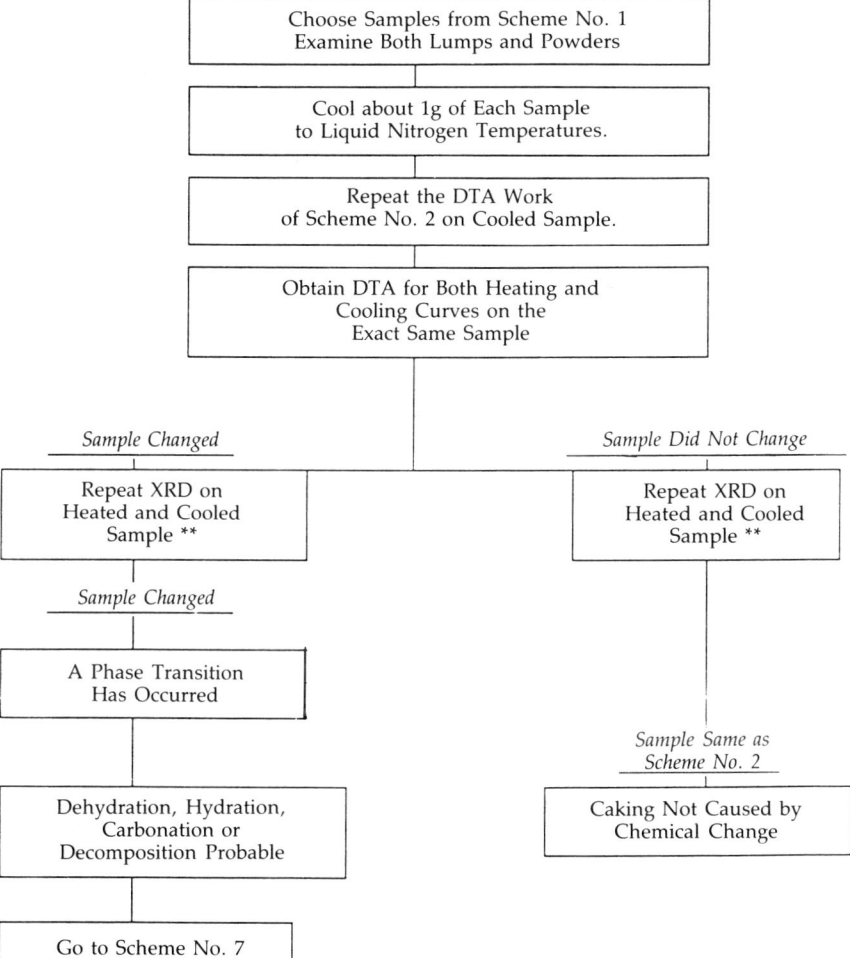

** *If DTA - TGA samples are too small for XRD heat and cool larger samples through the same temperature profile as the heating and cooling curves.*

SCHEME NO. 7

Solvates, Carbonates and Decomposition

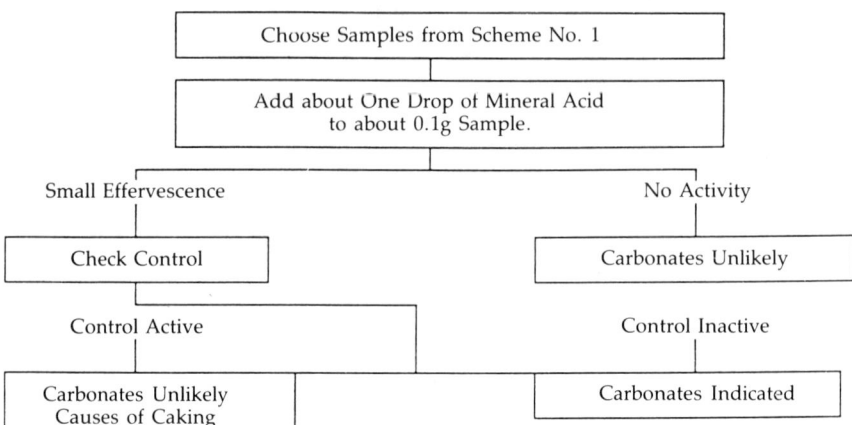

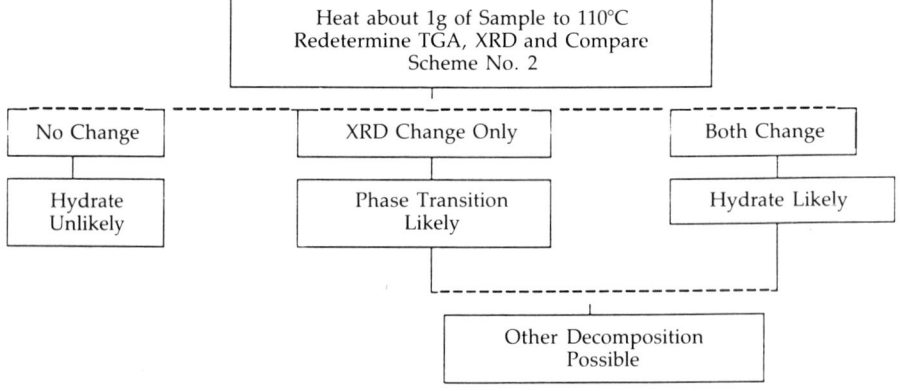

CAKING FLOW SCHEME NOTES

Notes for Scheme Number 1

Note 1. Establish that a caking problem does truly exist.
A. Has the solid been shipped in a wet or open container, exposed to weather?
B. Was the sample loaded into a container at an unreasonable temperature?
C. Is caking routine or sporadic?
D. Is caking seasonal?

Note 2.
Obtain several two pound samples from different lots, if possible. Observe the rules of gathering samples as outlined in the references of Chapter VI. Transport or ship the samples in tightly sealed thermos containers if available.

Note 3.
When samples are ready for study, transfer the samples to a clean, dry, cool (not cold) clear glass bottle. A small mouthed bottle is preferred if the sample can be transferred without modifying the sample. Small mouthed bottles usually seal more tightly than large mouthed bottles. If the sample must be broken up to be transferred, a wide mouthed bottle can be used. If a dry box is available, the transfers of hygroscopic materials should be made in the dry box but the sample should not be left open in the dry box. Hydrates can be lost as well as formed.

Note 4.
Obtain quality control analyses on the samples. Are the samples within all specifications other than caking? If they are not, the samples should be discarded for perhaps the wrong problem is being attacked. Obtain a second set of samples if they too are out of specification, then the raw materials, process, and the operators should be investigated thoroughly and quietly. Is this the source of the problem?

Note 5.
Store the samples where no great changes of temperature is likely to occur and at a place where direct sunlight is unlikely to fall upon the sample. Old samples should be discarded if they are older than the expected shelf life of a product.

Notes for Scheme Number 2

Note 6.
Scheme Number 2 is designed to obtain physical data to be compared with the control sample. Lumps are chosen because

the lumps are the most likely to differ from the well-behaved sample. If there is no differences that are evident from the physical measurements, chemical reaction mechanisms can probably be eliminated. The physical data will be used again in the schemes to follow.

Notes for Scheme Number 3

Note 7.
Mechanical caking is usually easily spotted once the sample is observed properly. Mechanical caking is usually associated with fibers and similar systems. Prilled products can be made that actually form hooks. It will depend upon the product as to whether anything can or should be attempted to prevent mechanical caking. Certainly, if some kind of barrier can be placed between the particles during manufacture, it may be possible to prevent the caking or wadding.

Notes for Scheme Number 4

Note 8.
Plastic flow is also easily spotted. Tars and waxes fall in this group. Particles are usually easily deformed but may cake only at temperatures above some critical temperature. The product may resemble a cluster of grapes where individual as spheres have flowed into each other. If the case is extreme, the particles may have coalesced into one single particle shaped by the container.

Notes for Scheme Number 5

Note 9.
The outcome of Scheme Number 5 and the other schemes to follow and much work is usually required just to classify the kind of caking that may be causing trouble. Static charges will be considered first.

Note 10.
After the samples have rested quiescently for at least twelve hours, gently turn the sample bottles on their sides. Note and record the response of each sample. Is the sample a free-flowing powder, a lumped powder, or one solid chunk?

Note 11.
Shake the bottles gently. Did any lumps disintegrate?

Note 12.
If the sample was a free-flowing powder and contained no lumps, swirl the sample in the bottle for about two or three minutes and then set the sample aside for thirty minutes.

Note 13.
If the sample is lumped and the lumps will disintegrate with soft shaking, then break up the lumps swirl the sample as in Note 12 and set the sample aside for thirty minutes. Because the particles return to their original condition when they are shaken and because of the speed that new lumps form, there is almost no mechanism that can be envisioned other than an electrical dipole interaction to explain the results when this behavior is observed.

Note 14.
Turn the bottles from Note 12 or Note 13 on their side again. Have the samples formed lumps again? If lumps have reformed, the caking problem probably has an electrical component as a contributor. In most cases the electrical caking will probably not form strong cakes but it is likely to position the crystals to assist in the formation of strong cakes if other kinds of caking forces become dominant.

Note 15.
If the sample in Note 10 was a lump too hard and strong to break by gentle shaking, remove a part or the lump for examination. In this examination the evidence of bridging between the particles will be explored. Bridging means that there has been mass transfer and the mechanism of the transfer must be isolated.

Note 16.
The test for the other electrical properties are described in detail in the body of this report. These properties are not obvious and are surely the basis of many caking problems that have persisted for years.

Notes for Scheme Number 6

Note 17.
Most caking problems are of the types studied in Scheme Number 6. It is the section where phase transitions of all types are found. Hydration, dehydration, solubilization, and recrystallization as well as the problems that can be caused by carbonization of basic systems or ammoniation of acidic systems can all be found in this section. An understanding of some sim-

ple physical chemistry will be required in this scheme. Most of the test are discussed in Chapter VI.

Notes for Scheme Number 7

Note 18.
Look at sample! If possible determine the type of bridging between the particles. Nothing else is as important as the first hand examination of a sample if it is the type of substance that can be handled without undue risk to the observer. Examine the feel of the sample by grinding the lumps between the fingers. Drop a lump on the floor or ground to observe what happens to the sample.

Note 19.
Look at sample with a polarizing microscope. Is the bridge between the particles crystalline? Is there a strain pattern between the particles? Often an amorphous material will exhibit a rainbow of color when viewed through polarized light. Glass blowers often observe strain in glass articles by examining the articles between large Polaroid lenses.

Note 20.
Add a drop of water to a few particles on a microscope slide. Is the sample readily soluble? Are the crystals hydrating? Can refractive index lines be seen streaming from the crystals? If the sample does dissolve, allow the solution to sit at room temperature to determine whether or not the water will evaporate and if it does what state the residue will have if the water will evaporate. Is it crystalline? Is it a film former? Is it sticky?

Note 21.
Heat a few particles of the sample to 110°C on a microscope slide. Did any change take place upon heating? Did a phase transition take place? Did the crystals disintegrate into many small crystals? It is probable that either a dehydration or a solid-solid phase transition has occurred if the above changes occur.

Note 22.
Weigh 5.0g of the sample into a small dish of known weight. Record the temperature and relative humidity. After the sample has been exposed for eight or more hours, reweigh the sample. Did the sample change weight? It should be determined what the nature of the weight change is before continuing. It is probably water but this is not certain until proved. It could be a volatile compound or a decomposition product. If it is proved to be

water, it should be determined how the water is contained in the sample. Is it free water? Is it dissolved water in an amorphous solid? Is it hydrated water? These questions can usually be answered easily with a thermobalance.

Note 23.
Depending on the outcome of Note 22 the sample should be subjected to either a higher or a lower vapor pressure to determine the rate of gain or loss of water under conditions the sample is likely to encounter. Most chemical handbooks describe the preparation of solutions that can be used in desiccators to precisely control the relative humidities of desiccator chambers. Determine the vapor tension at which the sample begins to gain or loose weight at ambient temperatures.

Note 24.
When it has been determined at what conditions the sample gains and looses, place a weighed, tared sample in the desiccator in which the sample just began to gain water and a second sample in the desiccator in which the sample had just begun to loose water. Weigh the samples as a function of time. Plot these data as weight versus time using simple graph paper. These plots will do much to characterize the sample and how it behaves toward water vapor. Do hydrates form? Does the solid deliquesce?

Note 25.
When a sample is removed from the desiccator, examine it carefully. Did it cake? Examine the sample under the microscope. Did the size or shape of the particles change while the sample was exposed to water vapor? Reweigh the samples and place them back into the desiccators but reversing the samples and desiccators. Now follow the loss of water by the samples that had been in the higher humidity while noting the gain of water by the sample which had been in the lower humidity chamber. Plot the data as before. Is there an inflection on either or both curves as the samples pass through their original weights? Reexamine the samples.

Note 26.
When this section is completed the cause or causes of caking should be evident. In some cases more than one type of caking may be causing a problem. An electrical set may be the beginning of other types of cake formation and the electrical set may only bring the particles close enough for other forces to take over.

SUMMARY

The flow schemes presented in this chapter are designed to lead the investigator through a series of tests that should indicate the type of generic problem encountered and what direction the project requires in order to properly access what should be done to solve the caking problem. In Chapter VIII case studies will be reviewed of selected solutions to caking problems and how the authors solved the problems. Many of the techniques presented here have been used by the authors but in no case were they organized in a "What to do next format." What has been done is very useful information in solving any new problem and the next chapter is a good starting point.

Notes to each section are included to help define what additional tests, if any, should be performed in order to attack the specific problem being approached. These notes are related to much of the information to be found in other sections and if more information is needed many of the references are directly related to the test. For these reasons no references will be included with Chapter VII.

CHAPTER
EIGHT

Typical Solutions to Caking Problems

COMMERCIAL PROBLEMS

In this work much attention has been given to the properties of solids as a means of studying and solving problems dealing with caking and flows of solids. Much more published work has been directed toward the design and manufacture of equipment to handle caked powders than to eliminate the problem at its source. Both approaches are necessary and some attention will be given to a small part of the literature describing equipment, its use, testing, and manufacture. A few references will be cited to give the reader a starting point in dealing with this area. Competition among the manufacturers is keen and the field changes rapidly as new and improved equipment is developed and manufactured. The capital investment in equipment of the type under consideration can be very great and considerable work and investigation usually precedes expenditures of this magnitude.

In dealing with commercial problems, ammonium nitrate probably ranks high on the list of substances that have been studied in great detail. Ammonium nitrate does just about everything to cause caking except form hydrates. It, therefore, has no water holding capacity. It is very soluble in even cold water. It undergoes several solid-solid phase transitions between room temperature and its melting point slightly below 170°C. It exhibits static electricity, piezoelectricity and pyroelectricity and at times cakes so badly as to require blasting to disintegrate large lumps. In 1921 a major disaster occurred in Oppau, Germany as a result of using explosives to break up a 4500 ton block of a mixture of ammonium sulfate and ammonium nitrate [90]. This blasting had been done many times before without troubles but on this morning the entire 4500 tons exploded. In reviewing the current literature dealing with ammonium nitrate the author is alarmed at the complete disregard for the hard lessons learned by others in the past while working with ammonium nitrate. Some of these hazards will be discussed in the section devoted to ammonium nitrate.

Foods also present an area where extensive work has been done on caking of everything from powdered avocado to cane sugar, table salt to starch [91] [92] [93] [94]. More work and a greater variety of solutions to caking of sugar have been reported than any other food product. Judging from the quantity of work still in progress none of the solutions seem to be completely satisfactory for all circumstances. Powdered milk or powdered eggs can also cause difficulties [95] [96]. Grains have received much attention [97]. In silos grains can create flow problems and in grain elevators dust and dust explosions are both hazards. Finished grains products such as flour, oatmeal, tapioca, and starch can cake [98].

The very stringent requirements placed upon all foods are very necessary and very limiting as to what can safely be added to a food product. The inorganic phosphates are outstanding in their safety record and are as safe as any functional substances known to mankind. When the phosphates will perform the required function, they are often a logical choice just because they are not only safe but are necessary nutrients for health and strength. An example is, tricalcium orthophosphate used as a flow conditioner in many products as is also fumed silica [99] [100]. The advantage of the phosphate is the fact that it is a nutrient while silica is inert.

The caking problems of table salt are ageless. Almost all processed foods contain some added table salt. The problems in dealing with most vital food products are many and no completely satisfactory solution may ever be found. The products of today are far superior to the products of the past, but one must admit that air conditioning and dehumidification have done as much to eliminate the caking of sodium chloride and similar products used in households. Other innovative science has contributed its share to the improvements. Even before central air conditioning became widely used, the home refrigerator played a significant role in preventing caking caused by biological attack or moisture. Microwave ovens have also done much to reduce the problems of caked ingredients used in cooking.

Brown sugar, which has been a worst case example of a difficult product to use, has the reputation of being a bad caker. Even badly caked brown sugar can be rendered free flowing by exposing it to microwave radiation for only a few seconds [101]. The microwave radiation is tuned to the frequency of water molecule absorption and the microwaves destructively crystalize the syrup bridging between the particles of sugar, allowing the sugar to be free flowing again.

Other products that shall be explored are fertilizers, cements and lime, coal, chemicals, plastics and fiber raw material stocks, dyes, ores, and clays. All of these products have caking and solids flow problems to greater or lesser degrees and all have been studied extensively. An attempt will be made to uncover the treatments that have been the most successful for both caking and flow problems but the major attention shall be directed toward caking. Many of the flow problems are related to equipment design and the flow conditioners may or may not help to prevent caking problems. It should be noted that engineers usually attempt to solve caking problems with mechanical devices while chemist trend to attempt process modifications and additives. Very often both approaches may be required. There are times when mechanical devices can make a system merely acceptable while the addition of a change in the properties of the product can make the product superior.

AMMONIUM NITRATE

The published literature dealing with specific examples of research done to understand or solve caking problems in products is extensive. In this chapter, examples that have yielded solutions to problems will be featured. Attention will also be directed toward those problems which still need a solution. For several reasons, ammonium nitrate will be chosen as the first material to be discussed. Not only are the chemistries and techniques interesting but some outstanding scientists have contributed to the work. Bridgeman, of high pressure work and diamond fame, was one of the significant contributors to the science [102]. It is interesting that Bridgeman used fuel oil as a continuous phase in his experiments with ammonium nitrate and seemed to have no problems with explosions. His life was charmed in many ways. The use of ammonium nitrate in explosives, rocket fuels, and fertilizers has contributed to the widespread interest in the chemical. It is also an oxidizer of choice when a bleaching agent is required in some industrial processes because it leaves no residue in the products it bleaches, particularly if the bleaching is done at temperatures in excess of 170°C.

The raw materials for the manufacture of ammonium nitrate are hydrogen, usually derived from natural gas, nitrogen, condensate water, and air. Ammonia is manufactured first from hydrogen and nitrogen which is allowed to react over an iron catalyst. The catalyst has usually been promoted with one or more metal oxides. A part of the ammonia is burned to nitrogen oxides, which are contacted with water to form nitric acid. The nitric acid and the remainder of the ammonia are allowed to react to form an extremely pure commercial product. Two factors contribute to the purity of the product. Most importantly many impurities cause the ammonium nitrate to become unstable toward explosion or decomposition and it can be hazardous to handle when impure. Secondly, when properly manufactured, there is little opportunity for impurities to enter the product. Most untreated commercial ammonium nitrate is more than 99.9% pure.

Ammonium nitrate melts at 169.6°C. It exists in four crystalline and one liquid form between room temperature and 170°C. The transition between the room temperature Form IV and Form III at 32°C is particularly troublesome if the product is stored. The transition temperature is in the summer ambient temperature range for much of Earth. A product is likely to undergo two transitions each day. There is a large density change at the transition plus the fact that mother crystals are converted to many daughter crystals which become mother crystals as the temperature of the product fall through the transition temperature when the product cools. This results in a swelling of the solid, causing it to cake and to even rupture bags if the process is continued.

Two varieties of ammonium nitrate are prilled while other products are formed as flakes. The prilled variety of the product is usually preferred. Prilling is similar process to the "shot tower" method by which lead shot were prepared during the Revolutionary War. The molten product is allowed to fall in a tower that is high enough to let the product form a sphere and freeze during transit. Surface tension of the melted product causes it to form nearly spherical beads as it falls but some additives as well as excessive ammonia in an ammonium nitrate melt can cause it to swell to "popcorn" type grains as it cools in the prilling tower. If the tower is too short or if the melt is too slow to freeze, the prills will splatter at the bottom of the tower.

The more common form of ammonium nitrate prills are manufactured from a melt containing about five percent water. The nitrate crystallizes as porous balls. It contains large open voids in the prills and can absorb large quantities of fuel oil without becoming oil damp. It is for this reason that it is preferred as blasting stock. The other form of ammonium nitrate is manufactured from an almost anhydrous melt. When prilled, the anhydrous melt yields a dense ball that has very little void space and absorbs fuel oil poorly. The dense form usually requires some kind of phase stabilizer in the nitrate to prevent the prills from suffering a phase transition while cooling to room temperature. The phase transition can completely destroy the prills leaving a powdered product that has little industrial or agricultural attraction but when the proper additives are chosen, magnesium nitrate being outstanding, the product can be stored for many years without caking problems provided it is protected from atmospheric water vapor.

The claims of methods to stabilize ammonium nitrate would fill a book much larger than this one and publications and patents continue to appear in the literature. Most of the methods do influence the caking characteristics of the product but most are of little practical value. A recent publication claimed diamine copper dinitrate as an anti-caking agent for propellant uses [103]. It may well be a very good anti-caking agent for its intended use but anyone with much knowledge of ammonium nitrate and its explosive history should never include an organic copper salt in large tonnage fertilizer type products. One commercial product includes borates in ammonium nitrate fertilizer. Borates are usually considered to be phytotoxic and it is interesting that they can be used in fertilizers. Dyes have been used to slow the rate of phase transitions in nitrate prills while one very successful industrial product that is still in production used magnesium nitrate to stabilize prills of ammonium nitrate [104]. Klyus et al. claim that the addition of ammonium polyphosphates will increase the transition temperature of ammonium nitrate from 42°C to 55°C. This is surely the case and the system is very interesting [105]. Ganz et al. used copper nitrate, magnesium nitrate, zinc nitrate, manganese sulfate, or sodium borate to raise the transition point of ammonium nitrate. They found that the zinc nitrate had the greatest influence. The author worked with all of these additives and again found after an exhaustive study that the magnesium nitrate was the superior system [106].

There have been no major disasters in recent years from ammonium nitrate explosions but the recent trend to forget the lessons of the past gives one pause for concern. It is the author's belief that most heavy metals and organic compounds should be meticulously avoided in the manufacturing and merchandizing of ammonium nitrate fertilizers. Sooner or later the practice can only lead to disaster, although it must be admitted that many years may pass before the disasters occur.

Many ammonium nitrate fertilizers depend upon a coating of clay to prevent the fertilizer from caking. Several clays can be used and usually one is about as good as another while the location of the plant with respect to the supply to clay will usually dictate the preferred clay. Ammonium nitrate, like most fertilizers, must be used within a zone of manufacture or the transportation cost will become prohibitive. In most cases the clay treatment is successful because very little ammonium nitrate is stored in the United States today. Most is handled in bulk shipments and most is used very soon after it is manufactured. This has many advantages. It is not stockpiled to present as much opportunity for caking or disasters. It is much less expensive to ship the product in bulk without the requirement of opening and disposing of sacks or other packages after the product is used.

Because so much of the current literature is filled with cases that have been exhaustively studied in the past, this subject will be treated in more detail than most other systems. Almost every system that had much of a chance of rendering ammonium nitrate non-caking was studied first from the standpoint of the thermodynamics of the phase transitions and in special temperature cycling cabinets to determine the long term influence of temperature cycling prilled ammonium nitrate.

Our work was directed toward anhydrous melts rather than to the more common water containing systems. A number of restrictions were imposed upon the additives which could be used in the anhydrous ammonium nitrate. Most heavy metals were excluded. Agronomic poisons as borates and boric acid were prohibited as were poisons such as arsenic and antimony compounds. Most organic compounds were prohibited in concentrations greater than a few parts per million, where some dyes and surfactants were effective, but these were added after the product exited the prilling tower.

When magnesium oxide is used in an anhydrous melt of ammonium nitrate, the resulting product is superior to any other form of ammonium nitrate ever manufactured when temperature stability, hardness, flowable non-caking properties are concerned. It has the added advantage that magnesium nitrate is a plant nutrient. Several companies in the United States built plants based upon this technology. Two approaches were taken to form the magnesium nitrate in the ammonium nitrate melt. In the initial laboratory work, magnesium oxide was added to molten ammonium nitrate and magnesium nitrate was formed in situ. When the first plant was built, magnesium oxide was dissolved in the nitric acid, which was then ammoniated to ammonium nitrate. When other companies made the same product, they chose to use the initial laboratory work and they made the magnesium nitrate in situ. The products are very similar but there is a slight preference for the addition of magnesium oxide directly to the anhydrous melt because the prills seem to be more stable.

The phrase diagram for the ammonium nitrate-magnesium nitrate system is presented in Chapter IV. One of the reasons that the prills are superior when made from the NH_4NO_3-$MgNO_3$ system results from the formation of a eutectic between the ammonium nitrate and the ammonium magnesium nitrate double salt, $(NH_4)_3Mg(NO_3)_5$. As is usually the case with eutectics, the crystallite size is very small and this leads to a very dense prill with a very high sheen to the surface. The prill is very hard. With most ammonium nitrate this would be highly undesirable if the ammonium nitrated suffered the normal 32°C phase transition. It can be noted from the diagram that when the concentration of MgO has reached 0.5% by weight the phase transition is over 50°C and this is well out of the ambient temperature range in the summer months in the United States. Unfortunately these prills are almost impervious to fuel oil and the product does not function well as a blasting agent.

Many attempts have been made to make the prills porous but the prills performed too well in their initial role as a non-caking fertilizer and they were never very good as blasting agents. The case is not true for rocket fuels however. In this case they can be made to perform very satisfactorily.

The two methods mentioned previously to keep ammonium nitrate from caking are (1) to bypass a transition by changing the systems response to thermal changes and (2) place a barrier between particles to prevent them from contacting and caking. A third method is based more on the kinetics of phase transitions than upon the thermodynamics of phase transitions. When certain dyes are dusted on ammonium nitrate, the phase transitions can still occur but the time required for the transition to take placed is so long the driving temperature may have returned to the lower value before the transition has had time to occur. Much work was done by Whetstone with acid magenta [107]. It worked very well but the red dye became difficult to contain in industrial processes, causing most of the plant equipment to be dyed red [108]. A near disaster occurred when dyed ammonium nitrate was being dispersed from a plane and no one had informed the pilot that the nitrate had been dyed. He saw the bright red flume jetting from his plane and mistakenly thought his plane was on fire.

FERTILIZERS

Fertilizers can range in properties from manures to reagent grade potassium phosphates. There are, of course, no caking problems with liquids or anhydrous ammonia. Specialty fertilizers can be small volume and often they are mixtures that are very complex with behavior that is difficult to predict even when one is supplied with a complete chemical analysis of the ingredients. Attention is directed to the large volume, conventional fertilizers such as ammonium sulfate, ammonium phosphate, urea, calcium nitrate, magnesium nitrate, calcium phosphates, potassium nitrate, sodium nitrate, and calcium carbonate and limes.

The fertilizers divide into two classes when viewed from the solubility and rate of solution properties. Some are very sparingly soluble and very slow to dissolve while the others are rapidly dissolved and may have a high solubility. Most of the ingredients are simple salts and have a simple chemistry as far as manufacture, caking, transportation, and application are concerned. The orthophosphates are not simple. Much of the chemistry of the pure salts is but poorly understood even in simple systems. In complex and impure mixtures and chemistry is even more complex. Since the phosphates are absolutely vital to any form of agriculture, their chemistry will be discussed in some detail in the section devoted to phosphate fertilizers and why caking problems can occur.

AMMONIUM PHOSPHATE

It has been claimed that basic aluminum salts would prohibit the caking of either ammonium phosphates or urea ammonium phosphates [109]. The action could be to react with the surface to yield an insoluble aluminum salt, or it could be merely the separation of particles by the more or less insoluble coating. Kuzko and Glabisz studied the impurities in the phosphoric acid made from so called wet phosphoric acid and how these impurities influenced the caking of crystalline ammonium orthophosphate made from the wet phosphoric acid [110]. They concluded that the lower the concentration of fluoride and the greater the concentration of aluminum the less inclined the ammonium phosphate is to cake. This agrees with the claims of Borisov et al. in the previous reference [109].

Wet phosphoric acid is the acid made by treating phosphate rock with sulfuric acid and then filtering the calcium sulfate that is formed from the acid. Even purified wet phosphoric acid is not as reliable in purity as phosphoric acid made from elemental phosphorus. Distillation is a superior unit process to solvent extraction with respect to purification. Several companies have recently introduced purified wet acid into the phosphoric acid market of the United States. It will be interesting to observe how well they compete with furnace phosphoric acid that has had a long and very enviable record for purity in the food phosphate markets.

AMMONIUM SULFATE

Ammonium sulfate has been used as a fertilizer for many years but has a rather low nutrient value on a per pound basis. It also will cause the soil to become acidic if the sulfate ion is not treated with lime or a similar basic substance. The Russian school continues to publish about the caking characteristics of ammonium sulfate and methods of preventing it from caking. Amines are reported to be very helpful and unlike ammonium nitrate no problems can occur that might lead to explosions since the sulfate ion is a much weaker oxidizing agent than the nitrate ion [111]. Acrylic polymers are also reported to give very good results in preventing caking even under high vapor tension loading [112]. In another study it is concluded that water is the chief cause of ammonium sulfate cakes under load conditions [113]. This is probably a valid conclusion but many other contributing factors can be excluded in such a general conclusion.

UREA

Urea is very soluble in water and most of its caking problems can be directly related to its high solubility. An interesting account of the use of oxamide (oxalic diamide) as a anti-caking agent for urea has been reported [114]. It is claimed that urea containing 2.5% or more oxamide is practically noncaking. Oxamide is very sparingly soluble in water while urea is very soluble. Since no phase diagram of the system had been published, it was decided that the aqueous phase diagram of oxamide and urea would be determined especially for this book. The aqueous three-component system, water-urea-oxamide phase diagram is presented in Chapter IV. As seen in the phase diagram, the oxamide is very sparingly soluble in both water and urea and is acting as an insoluble flow conditioner much as a clay might have behaved to prevent caking.

As a typical solution to solving caking problems, several other factors should be recognized in dealing with these phase systems. The phase diagram is needed over and above merely recognizing that oxamide is effective in reducing the caking tendencies of urea. The phase diagram offers clear options concerning how much of a second component is needed to establish a required resistance to hygroscopicity. Better estimations of the physical chemical behavior of the solutions can be calculated for neutral molecules than can be done with ionized salts, as demonstrated in Chapter II, where vapor pressure lowering was discussed.

The oxamide-urea diagram should be compared with the carboxylate detergent builder-sodium carbonate phase diagram in Chapter IV in which it was noted that sodium carbonate reduced the hygroscopicity of the detergent builder, but it was not as effective as desired because the sodium carbonate is almost as soluble as the builder and the system of solid solutions does not extend all of the way across the system but breaks into an equilibrium between the solid solution and sodium carbonate hydrates.

Although this is not the case, it should be expected that the urea-oxamide system should form a solid solution. It can be anticipated that the two compounds with low molecular weights and similar structures might form crystals that are either solid solutions or double salts.

$$\underset{\text{urea}}{\text{H}_2\text{-N-}\overset{\text{O}}{\text{C}}\text{-N-H}_2} \qquad \underset{\text{oxamide}}{\text{H}_2\text{-N-}\overset{\text{O}}{\text{C}}\text{-}\overset{\text{O}}{\text{C}}\text{-N-H}_2}$$

The urea-oxamide system points up again the great difficulty that can be encountered in predicting the solubility of a salt. If one considers how similar urea and oxamide are, differing by a single carbonyl group, it should be expected that their solubility in water should be similar. Urea has a specific gravity of 1.335 g/cc at 20°C while oxamide has a specific gravity of 1.667 g/cc. The molar volume of urea is 44.99 cc/mole while the molar volume of oxamide is 52.83 cc/mole. This represents a differential volume 7.9 cc/mole of carbonyl groups and is a small number, indicating that the molecules are tightly packed. The solubility of urea is 100 g/100 cc of water at 17°C while oxamide is less than 0.1 g/100 cc at the same temperature. Urea is more than 1000 time as soluble as oxamide!

There have been a number of very interesting articles and patents dealing with the properties of oxamide. The work was probably done more because the compound has been used as a burning modifier in rocket fuels than because it was used in fertilizers but the information is very useful in either case. Even the method of synthesis is very interesting. Two molecules of formamide are coupled in a glow discharge to form the molecule.

$$\text{H}_2\text{-N-}\overset{\text{O}}{\text{C}}\text{-H} + \text{H-}\overset{\text{O}}{\text{C}}\text{-N-H}_2 \Rightarrow \text{H}_2\text{-N-}\overset{\text{O}}{\text{C}}\text{-}\overset{\text{O}}{\text{C}}\text{-N-H}_2 + \text{H}_2 \qquad [18]$$

Since urea and ammonium nitrate are both used as fertilizers, another precaution that should be mentioned resulted from a recent conservation with an inquirer requesting information. The caller was mixing urea and ammonium nitrate and was surprised that his plant was suffering small detonations that were blowing pipes apart. It is well known that urea and ammonium nitrate mixtures are very explosive as is urea nitrate. Urea acts as a fuel for the reaction in a manner similar to fuel oil in the fuel oil ammonium nitrate blasting agents. It should also be recognized that only 6% by weight fuel oil is required to convert the fertilizer into a blasting agent. A similar weight of urea mixed with ammonium nitrate could cause an entire plant to detonate. I cannot overly stress the need to be very familiar with the older literature before recommending changes in processes that could lead to either explosions or poisons. The responsible scientist knows the published literature and company reports dealing with the product he or she is working. *It Should Be Obvious That Oxamide Should not Be Mixed With Ammonium Nitrate Without Exercising Extreme Care!* Oxamide would probably improve the caking characteristics of ammonium nitrate but the risk of an explosion cannot be minimized even if the end use of the ammonium nitrate is in a rocket fuel where it has been reported to have been used.

PHOSPHATE FERTILIZERS

The problem with agricultural phosphates is, and always has been, keeping them soluble long enough for the phosphates to be available to plants. This is true whether the plants are corn in a field in the midwestern United States or are the aquatic plant food in a fish farm in China. It has long been known that phosphates must be added to the ponds in which catfish are raised if enough fish are to be produced to make the operation of benefit to mankind.

When phosphates are mined for fertilizers, we are very often injecting ourselves into a food chain that died between 100 million and 600 million years ago. Much of the phosphates are being recycled and have been in life forms at least once before, much as fossil fuels have all been in a life form at some time in the past. The fact that some of the phosphates are still intact as bones and shells after millions of years attest to their very low solubilities and slow rates of leaching but they are leached and may be reabsorbed by limestone deposits.

The phosphates that are mined or the bone meal from animal bones can be used as a source of phosphates for fertilizers but they are not satisfactory for a hungry world. They are not soluble enough to furnish much plant food in the growing season of a plant and for this reason the phosphates are treated with acids to make them soluble enough to feed the plants. The peoples of the Earth cannot wait until next year to eat. Even with the acidified phosphates if the plant cannot use the phosphate this growing season by next growing season most of the phosphate has become too sparingly soluble to be of much use. Again there is no truly satisfactory solution to this problem and the farmers are forced to add new active fertilizers each growing season if the plants are to be fed. Ideally a controlled release fertilizer which supplied nutrients to the plant at an optimum rate should be preferred. Perhaps some biological approach to this problem might result in less phosphorus being mined each year. An organism capable of attacking apatite and other insoluble phosphates each spring might be capable of solubilizing enough phosphate to feed plants during a growing season when the soil is damp and warm.

Hopefully this background will prove beneficial in the discussions of the chemistry of caking of phosphate fertilizers. There are very many forms of calcium orthophosphates and double salts with both anions and cations. To keep things simple, consider only three of the calcium orthophosphates and their hydrates. As mentioned the two salts, $Ca_5(PO_4)_3F$ and $Ca_5(PO_4)_3OH$ are the starting place. In good grades of fertilizers the ore is first defluorinated after it has been subjected to a flotation process to remove sand, metals, and so forth that would act as impurities. Defluorination can be done by heating the $Ca_5(PO_4)_3F$ in a calciner. The purified "phosphate rock" is then treated with either sulfuric acid to make super phosphate fertilizers or with a crude grade of phosphoric acid that was itself made by treating the phosphate rock with sulfuric acid. This technology has improved over the years but the article by H. M. Stevens in *Phosphorus and Its Compounds* is still an excellent background [115].

The two calcium phosphates considered as derived from the apatites, $Ca_5(PO_4)_3F$ and $Ca_5(PO_4)_3OH$, are $Ca(H_2PO_4)_2$ and $CaHPO_4$. Dicalcium orthophosphate, $CaHPO_4$, is almost as low in solubility as $Ca_5(PO_4)_3F$ and if the $Ca_5(PO_4)_3F$ were acidified to $CaHPO_4$ there should not be much improvement in the availability of phosphorus to a growing plant. But monocalcium orthophosphate, $Ca(H_2PO_4)_2$, is very soluble and can supply dissolved phosphate to the roots of a plant. (Unless the phosphate is dissolved very little phosphorus can be transported into the plant.) Now there is another problem. $Ca(H_2PO_4)_2$ is thermodynamically unstable and disproportionates as in the following equation.

$$Ca(H_2PO_4)_2 => CaHPO_4 + H_3PO_4 \qquad [19]$$

It is not difficult to understand why a phase transition of the type outlined previously should cause a caking problem. A new crystal form and phosphoric acid are released if the transition occurs.

Most manufactures of technical grades of $Ca(H_2PO_4)_2$ over neutralize the salt with either lime or some more soluble base. Lime is the least expensive choice. This makes a free flowing powder or granule but a part of the increase in solubility has already been lost and the increase in pH will assist in the precipitation of more of the phosphate with the metals of the soil. Again there is no truly ideal answer.

In most cases one would expect $CaHPO_4$ formed in the previous equation to be ineffective as a source of acidity because the dissociation constant for the hydrogen ion is so small. The hydrogen cannot even be titrated directly with the usual sodium hydroxide analytical reagents. Dicalcium orthophosphate dihydrate is used as a leavening agent in baked goods. A leavening agent is an acid that can react with sodium bicarbonate to release carbon dioxide in bread and other bakery items to make them rise and be light in weight and texture. The reason that $CaHPO_4$ can be used is a result of tricalcium orthophosphate or hydroxyl apatite being so sparingly soluble. Because of this, dicalcium orthophosphate also disproportionates to monocalcium orthophosphate as in the previous equations and finally all of the phosphate is converted to tricalcium orthophosphate and orthophosphoric acid. The orthophosphoric acid reacts with the sodium bicarbonate to yield sodium orthophosphate, carbon dioxide, and water of course. To simplify the following equations, the equations will be written as though tricalcium orthophosphate formed when in reality it is apatite.

$$4CaHPO_4 => CA(H_2PO_4)_2 + Ca_3(PO_4)_2 \qquad [20]$$

And as above

$$Ca(H_2PO_4)_2 => CaHPO_4 + H_3PO_4 \qquad [21]$$

until the reaction consumes all of the acidity.

DETERGENTS

Throughout the world much work is still devoted to the caking of phosphate detergents. In Europe the combination of perborates and phosphate combinations has been studied at length. Gun'ko et al. concluded that organics as surfactants, fatty acid soaps, monoethanolamine, and the like caused detergents to flow more poorly, while inorganic compounds as sodium tripolyphosphate, sodium metaborate, sodium carbonate or sodium sulfate caused the detergents to dissolve more slowly and show less tendency to cake [116]. Tsutazumi, et al. arrived at similar conclusions but in both cases the formulations are so complex as to make it difficult to determine the role of separate ingredients [117].

Garner-Gray, et al. studied sodium perborate formulations and derived an equation related to caking they called the perborate caking index, PCI.

$$PCI = A + 31.25 \, V - 1625 \qquad [22]$$

where:
A = the surface area of the detergent in m_2/g
V = pore volume in ml/g

If PCI was less than zero, the detergent was reported to free of caking but when surfactants were added bleaching with the perborate was satisfactory only if the PCI was greater than zero. There is a trade-off

198 / Typical Solutions to Caking Problems

between quick bleach delivery and caking tendencies [118]. York and Atkinson claim that phosphate esters and a hydrotrope sodium toluenesulfonate aid in the preparation of non-caking detergents. Their system contained 24% sodium tripolyphosphate and this does much to reduce the caking of detergents [119].

Borrello claims that the degree of hydration of sodium perborate in a detergent mix can influence its caking and that the lower levels of hydration are superior to hydrations as great as sodium perborate tetrahydrate [120]. There is small doubt that this is true. As noted in Chapter III, the greater the degree of hydration the more likely the salts are to release their waters of hydration at lower temperatures. The freer the water in any formulation the more likely it is to cake.

Clarke et al. claimed that a detergent mix containing sodium tripolyphosphate and petroleum jelly gave superior flow and caking properties [121]. Bouillet, et al. claim that a magnesium silicate and alkyl aryl sulfonate form an anti-caking agent and prevented agglomeration [122]. Yamamoto et al. claim that silicic acid and magnesium carbonate added to hydrated perborate will prevent the perborate from caking [123].

It is known, for example, that if sodium tripolyphosphate hexahydrate is coated with urea and the sodium tripolyphosphate is dehydrated, the integrity of the anion is not violated, simply because the surface was protected against hydrolytic degradation [132]. This protection is not a result of any type of encapsulation but rather that urea is a dehydrating agent that will react with water to yield carbon dioxide and ammonia.

$$H_2\text{-N-}\overset{\overset{\displaystyle O}{\|}}{C}\text{-N-}H_2 + H_2O \Rightarrow CO_2 + 2NH_3 \qquad [23]$$

Water is converted to two gases and is lost from the system, leaving an anhydrous product behind. Additionally, it will convert acids to anhydrides or in the case of phosphates cause the formation of POP linkages from acid constituents. Without urea water will hydrolyze POP linkages, degrading the polyanion.

FOODS

There are many food products which have caking problems. One area that is probably not encountered by most interested readers is in vending machines where powdered coffee, powdered tea, powdered cream, and even powdered soups are dispensed. These powders must be free flowing for several reasons. Sanitation is high on the list.

If the products become moist and lumped, bacterial and fungal growth is encouraged. The powders must be dispensed in measured quantities, otherwise the drink or food will be too concentrated or too dilute when delivered. If the caking becomes severe, then only water is delivered. These problems have been approached both by food chemistry and by equipment [124]. The modern vending machine is ubiquitous and so sophisticated it is often taken for granted. Some of the machines not only make change for paper bills, they even talk and answer questions.

SUGAR

Sugar has received much attention because of its caking behavior and the extremely large market. Because sugar is very soluble in water (211.4 g/100 g water at 25°C), very hygroscopic, particularly when freshly ground, and forms no hydrates it is not surprising that it cakes. Many of the approaches that can be used in industrial systems are not permitted in a food product. Sugar is also subject to infection by micro-organisms, further limiting its use in some systems.

It is not too surprising that it has been reported that the surface layers of sugar are more hygroscopic than the interior layers [125]. It is known that the surface of sugar becomes amorphous when sugar is crushed and it will very quickly absorb water from the atmosphere [126]. But when the vapor tension of water in the atmosphere is great enough, the sugar will absorb enough water to allow the surface sugar to recrystallize and in so doing the sugar will release water back to the atmosphere. This in turn causes the sugar to lump and cake badly as a result of the phase transition. It is claimed that spraying the freshly ground sugar with hydrated peanut oil will prevent it from caking [127].

Others claim that merely lowering the water content of sugar below 0.03% before it is milled will produce a sugar that will not cake [128]. Nisshin Seito claimed that the addition of maltose to sugar will prevent it from caking [129]. It is also claimed that passing sugar through a fluidized bed just after milling it will prevent it from caking [130]. Fludizing probably allows the amorphous phase to recrystallize under conditions that will not allow the particles to contact and bridge. A similar treatment was recommended by Chapman [131].

There is one very important point of emphasis made in most of the work on sugar. If the surface of a product can be controlled, very often it will control the entire system. This means that a very small quantity of a substance on the surface of a powder may be more helpful than a larger quantity in the particles themselves.

The literature devoted to sugar is so extensive that hundreds of articles could be cited. The ones presented here are intended to be only a sampling and to give the reader a feeling for the type work that has been done in this most important food area. So much has been reported it is difficult to determine which treatments are practical and which are not. One thing is very clear. The ideal solution to the caking of sugar is yet to be discovered. Perhaps the wide variety of uses prohibits any ideal solution to satisfy all conditions.

SALT

Salt has been the subject of endless crystallization studies directed toward reducing its caking tendencies and changing its crystal habit. Some commercial table salt is crystallized as octahedral particles by adding a small quantity of urea to the crystallizer. Caking has been attributed to trace quantities of calcium and magnesium salts in most commercial sodium chloride. The source of the salt would have a strong influence on the impurities since most salt is derived from solar evaporation of sea water or is mined. Many flow conditioners have been suggested, including magnesium carbonate, sodium ferrocyanide, tricalcium orthophosphate, fumed silica, and many others. High humidities combined with high temperatures will probably always cause unprotected table salt to cake. The solubility of the salt is the major property that must be controlled and it is unlikely that a very small concentration of any flow conditioner can be expected to perform under tropic conditions.

A fundamental approach to the problem has the best chance of yielding lasting results in the caking of sodium chloride. If a solid solution of sodium chloride and a second less soluble salt can be found and the phase diagram tie lines exhibit the proper slope, it is possible that a lasting solution can be found. Likely candidates should have crystal lattices that are similar to sodium chloride and are known to have much lower solubilities. It is sometimes stated that such crystals are isomorphous. It may or may not be necessary that the system of solid solutions be continuous form one pure salt to the other in a three-component aqueous phase to be effective but the best results for small additions of the second salt will occur when the series of solid solutions is continuous.

One of the very obvious problems in dealing with sodium chloride additives is the possible toxicity of the added component. This is particularly true if the salt is to be used in food products. The aqueous phase diagrams of substance such as silver chloride or maybe thallous chloride, TlCl, should be determined as pilot studies. The salts are too toxic for food uses. Both salts have low solubilities but silver chloride is soluble in sodium chloride solutions and this was at one time used as a method of fixing photographic plates before sodium thiosulfate was known. Both salts have cubic crystal lattices. Both salts have a chance to form solid solutions with sodium chloride but both salts discolor in light or air. This may or may not be a problem in small quantities in sodium chloride. Both salts are very expensive compared to sodium chloride. These would be interesting pilot experiments but it is unlikely that they could be made practical for most uses. They are merely suggested as examples of how the problems can be approached and are not intended to be presented as proved solutions.

A second problem is psychological. It is often difficult to gain support to add a substance to a product if the added substance is much more expensive than current product. Very few substances are much less expensive than sodium chloride.

Several references will be included as starting points for studies on the caking of sodium chloride. None of the works claim complete victories. Wiedorn used drying agents in containers where salt was shipped or stored to prevent caking [103]. Meyer, Kesler, Richards, and Robe report that sodium ferrocyanide is outstanding below 70% relative humidities [133]. Today this approach should be carefully studied from a toxicology point of view before the salt is considered for food use because hydrogen cyanide may be generated when sodium ferrocyanide is exposed to light and strong acids and bases can also cause decomposition [134]. Jakinovich obtained a patent for coating sodium chloride with a saliva insoluble zinc salt which is claimed to be non-caking [135]. Bueckenhueskes and Gierschner described the properties of salt according to source as sea salt, rock salt, and so forth with particular interest in the canning industry and how caking could be eliminated under use conditions [136]. A book could easily be written about the caking properties of sodium chloride alone without using but a small portion of all that could be included.

SWIMMING POOL CHEMICALS

As a result of the large number of homes in the United Sates that are equipped with backyard swimming pools, the market in these chemicals is large and competitive. Most of the chemicals are of a type that are inclined to cake. The common oxidizing agents used to disinfect the pools are of the same type used in bleaches for fabrics and home laundry applications. They range from sodium hypochlorite, which is used as a liquid to calcium hypochlorite (bleaching powder) or chlorinated cyanurates. Chlorinated trisodium orthophosphate has also been used as has sodium perborate and other peroxides. All of these chemicals are unstable and are said to lose available chlorine or available oxygen. It has been learned that when a chemical decomposes it undergoes a phase transition. Phase transitions are inclined to cause crystals to cake.

Swimming pool chemicals are used and stored in damp or humid conditions around a pool in most cases. Most of the degradation products of the bleaches are basic, and as a consequence, they react with carbon dioxide of the atmosphere to form carbonates. The hydroxides in the system are converted to water and the corresponding carbonate. Additional phase transitions occur at each reaction. The obvious first choice is to take meticulous care in the manufacture of swimming pool chemicals to insure that they are dry when packaged. The second care is to package the products in a container that is both water and gas proof. If the customer is properly warned to keep the chemicals well sealed, the chances are the products can be used without undue degradation provided the shelf life is realistically obeyed.

An excellent booklet, *Everything You Always Wanted to Know About Pool Care,* was published by Charlie Taylor [137]. In the booklet he discusses not only the problems with di- and trichlorinated cyanurates but also the stabilization of hypochlorites with cyanuric acid. The concepts of calcium carbonate formation from calcium hypochlorite is covered in detail as well as the caking of sand filters. Three types of filters are used in home pools. One filter is sand, one is diatomaceous earth filter, and the other is a cartridge type that may be filled with any of several substances but if it cakes it is thrown away a new cartridge is installed. With the other two filters it is possible to clean them as they are backwashed using a sequestering agent of a polyphosphate or ethylene diamine tetraacetic acid (EDTA). For those interested in swimming pool care, the book is well worth reading and although it does not address caking per se, it does give many clues about the properties of the chemicals that are instructive to one who has mastered the materials covered. It should be interesting to obtain this or a similar book and attempt to determine which of the chemicals can be expected to experience caking problems when the properties of the chemicals are understood.

Much of the technology associated with swimming pool chemicals is directed toward agglomeration and induced cake formation. For this reason much of the chemistry can be found in Chapter IX where it seems to be a better fit.

WATER TREATMENT CHEMICALS

The caking of sand filters was mentioned with respect to swimming pool applications. Of even greater interest is the caking of sand filters in the purification of potable waters and water in treatment chemicals in general. One of the prime reasons water treatment companies have sand filters in their process is as a final polishing of the water to insure the public that the water is of the highest possible quality.

The reason that sand filters become unusable is a result of scaling or cementing with calcium and magnesium carbonates and other salts that contribute to water hardness. In some parts of the United States the filters can become unusable in as little as three weeks if the water is not treated with some kind of threshold agent. A threshold agent is a substance that prevents the precipitation of scales when the threshold agent is used at very small concentrations. About 100 ppm is typical and sodium polyphosphate glasses, "Sodium Hexametaphosphate", are the agents of choice because they are both effective and very safe to use. Because the sodium polyphosphate glasses are amorphous and very hygroscopic they are inclined to cake. The caking tendency can be reduced by mixing the glasses with sodium carbonate powder. With proper treatment a water company sand filter may last for several months rather than a week or two. Additionally, the water heater in homes last much longer when the potable waters have been treated with polyphosphates because scaling is reduced.

The phosphate glasses have been used for many years in the treatment of red water and green water in home and industrial uses. Green water is usually caused by copper while red water is usually iron staining. In order to lessen the caking problems with sodium phosphate, glasses the incorporation of calcium salts in the glass will lower the rate of solution of the salt and lessen its caking tendencies.

To get a first hand accounting of the status of the caking of chemicals used in potable water systems, the director of one of the major water companies was consulted. Ferrous sulfate is the only chemical used that is likely to cake other than the polyphosphate glasses mentioned above. When ferrous sulfate becomes hygroscopic it also becomes acidic and aggressive toward metals. If ferrous sulfate cakes very badly, it can be replaced. The sodium polyphosphates are used in such low concentrations that the lumping causes no major problems and can be easily dissolved in warm water.

METALS TREATING

A wide variety of solids are used in metal treating and the some of the products cake before the are used if they are retained too long. Soft steel is treated with an anti-corrosion agent often referred to Jernstedt's salts. It is chiefly a disodium orthophosphate with some colloidal titanium salts that activate the metal to causing the metal to coat with zinc orthophosphate. The salt is usually made by causing the disodium orthophosphate to cake and then crushing or grinding the material before it is packaged. If the salt is not properly cured or if the salt is left exposed to very humid atmospheres as are often found in metal treating shops, it is likely to cake. The only known cure for the problem is to manufacture the product correctly and keep it properly contained after it has once been opened to the atmosphere.

The entire metal treating and paint industry is involved with powdered solids such as pigments, deflocculating agents, fungicides, dyes and wetting agents. Most of the pigments are extremely sparingly soluble in the liquids with which they are mixed and caking problems are usually confined to the other constituents of the paint. Sodium tripolyphosphate is one of the preferred deflocculating agents used in water based paints and it is usually free of caking problems provided it has been properly manufactured. If the sodium tripolyphosphate is not manufactured correctly, it is almost certain to cake and the best solution to the problem is to find a new supplier if caking is a common occurrence. Trisodium orthophosphate is used to clean many surfaces that are prepared for painting. Trisodium orthophosphate is a strange and interesting compound. It is seldom, if ever, Na_3PO_4 but usually is a double salt of sodium hydroxide or sodium carbonate. This will depend upon the manufacturer and the raw materials used to prepare the salt. If sodium hydroxide is used to neutralize the phosphoric acid, it is likely that the compound could be $Na_3PO_4 \cdot \frac{1}{4}NaOH$ or similar. The $\frac{1}{4}NaOH$ may vary considerably with respect to the phosphate. The greater the quantity of NaOH in the *crystal* the higher the pH, the more likely the compound is to suffer a phase reaction to sodium carbon dioxide, and the more likely the salt is to cake. Nevertheless, trisodium orthophosphate is not inclined to cake and is used in hard surface cleaners used in the household, probably as a result of its low solubility in water.

Much sodium hydroxide and some potassium hydroxide is used in the metal treating industry. Most sodium hydroxide is sold and shipped as 50% solutions. Some is sold as a gigantic cake made by pouring molten caustic soda into drums that contain an average of 363 kg. The sodium hydroxide is removed from the drum with huge can openers in much the same way phosphorus pentasulfide is handled. The flaked material is used in the process as quickly as possible, thus avoiding the caking problems that are certain to occur with materials as hygroscopic and reactive as sodium hydroxide [138].

The subject of metal treating and cleaning, both ferrous and nonferrous, is covered as an overview in *Chemical Technology* [139]. No particular attention given to the subject of caking but it can be seen from the list of substances used that some are known to cake.

SUMMARY

Caking problems occur in one form or another in almost all powdered products. In some areas the same substance will be studied over and over with solution after solution being given for a caking problem. The problem is solved for a given application to a customers satisfaction but when the same ingredient is applied in another product it may not satisfy the customer and the study is repeated for yet another set of conditions.

In some cases the problem has been solved but there are so many counterclaims that the investigator facing the problem anew is not certain which of the claims to respect and tries yet another approach. The attempt has been made to present some of the classic studies without specific recommendations as to which are the best for any given set of conditions. With only the literature and personal experience to depend on it should be risky indeed to pass judgement in specific areas where the author has had little or no experience.

REFERENCES

90. Kast, H., *Z. Ges. Schiess Sprengstoffw 20*, 43 (1925).
91. Santos, S.C.S., and Cal-Vidal, J., *Pesqui. Argop. Brasil 20*, 615 (1985).
92. Stengl, R., *Listy Cukrovarnicke 100*, 140 (1984).
93. Anon., *Food Eng. 41*, 116 (1969).
94. Peleg, M., Mannheim, C.H., and Passy, N., *J. Food Sic. 38*, 959 (1973).
95. Warburton, S., and Pixton, S.W., *Dairy Indus. International 43*, 23 (1978).
96. Lai, C.C., Gilbert, S.G., and Manheim, C.H., *J. Food Eng. 5*, 321 (1986).
97. Takagi, F. and Minami, K., *J. Japanese Soc. Ag. Mach. 48*, 349 (1986).
98. Mannheim, C.H., and Passy, N., *Leben. Wissen. Tech. 15*, 216 (1982).
99. Bauder, U., *Food Eng. Int. 3*, 23 (1978).
100. LaBell, F., *Food Proc. 47*, 80 (1986).
101. Borton, M.J., Private communication, Monsanto Company (1989).
102. Bridgeman, P.W., *Proc. Am. Acad. Arts Sci. 51*, 581 (1916).
103. Engel, W., *Propellants, Explosives, and Pyrotechnics 10*, 84 (1985).
104. Griffith, E.J., *J. Chem. Eng. Data 8*, 22 (1963).
105. Klyus, I.P., Skrypnik, I.G., Brezgin, B.N., and Bartosevich, R.D., *J. Appl. Chem. USSR 57*, 1702 (1985).
106. Ganz, S.N., Varivoda, I.K.H., Kuznetsov, I.E., Dinkevich, I.D., and Larina, L.M., *J. Appl. Chem. USSR 43*, 732 (1970).

107. Whetstone, J., *Trans. Faraday Soc. 51,* 973, 1142 (1955). *J. Chem. Soc.,* 4841 (1956). *Acta. Cryst. 7,* 697 (1954).
108. Buchart, A., Whetstone, J., *Disc. Faraday Soc. 1949,* p. 254.
109. Borisov, V.M., Azhikina, Yu.V., and Gerke, L.S., *Khim. Promyshlennost 8,* 590 (1978).
110. Kuzko, A. and Glabisz, U., *Przem. Chem. 55,* 211 (1976).
111. Rassadin, B.V., Sorokina, N.P., Kuzentsova, V.V., Tsekhanskaya, Yu.V., *Zhurnal Prikladnoi Khimii 59,* 1299 (1986).
112. Kuznetsova, N.S., Kruchinina, N.D., Groshev, G.L., Rybin, G.V., Ermilin, Yu.N., Kozodoi, V.M., Aurov, A.P., *Khimicheskaya Promyshlennost 10,* 46 (1976).
113. Vorob'er, S.E., Kurinnaya, S.N., Kolesnikova, V.I., *Koks i Khimiya 8,* 38 (1986).
114. Parkhomenko, V.D., Smirnova, E.S., Pivovarov, A.A., and Steba, V.K., *Khim. Tekh. (Kiev) 4 (142),* 7 (1985).
115. Stevens, H.M., *Phosphorus and Its Compounds Vol. 2,* ed., Van Wazer, J.R., Interscience Publishing Company, New York, N.Y., p. 1025 (1957).
116. Gun'ko, A.P., Stadnik, V.F., Shkarovskii, I.A., and Zabrodskaya, T.N., *Kihm. Tekhnol. (Kiev), 6,* 26 (1988).
117. Tsutazumi, J., Obara, M., and Iguchi, K., Japan Patent No. 63196698 A2 (August 1988).
118. Garner-Gray, P.F., F'Arnworth, P., Parslow, M.W., and Sims, P.S., Europe Patent No. 164778 A1 (December 1985).
119. York, D.W. and Atkinson, N.J., British Patent No. 2158087 A1 (November 1985).
120. Borrello, G., German Patent No. 3509331 (March 1985).
121. Clarke, K., Davies, R.L. and Nicholls, D., European Patent No. 30859 (June 1981).
122. Bouillet, E., Logan, W.R., and Sarot, P., British Patent No. 2018843 (October 1979).
123. Yamamoto, S., Zonezawa, M., and Kinishima, T., Japanese Patent No. 52012159 (April 1977).
124. Schubert, H., *J. Food Eng. 6,* 83 (1987). Ehlermann, D.A.E., *Reliable Flow of Particulate Solids,* Chr. Michelson Inst., Bergen Norway, p. 7 (1985).
125. Kelley, F.H.C., Mak, F.K., and Shah, D., *International Sugar Journal, 76,* 361 (1974).
126. Roth, D., *Zucker, 30,* 274 (1977).
127. Baum, W., Gierlichs, M., West German Patent No. 1,492,837 (1969).
128. Gawrych, M., *Gezata Curownicza, 77,* 57 (1969).
129. Nisshin Seito, K.K., Japanese Patent No. 5,646,760 (1981).
130. Ashlers, H., et al., East German Patent No. 2,028,470 (1971).
131. Chapman, F.M., *Zeitschrift Fuer Die Zuckerindustrie, 21,* 12 (1971).
132. Griffith, E.J., *Pure Appl. Chem. 44,* 184 (1975).
133. Meyer, D.W., Kesler, W.M., Richards, T.H., and Robe, K., *Food Processing 29,* 49 (1968).

134. *Merck Index*, Ed. 8, Merck and Company, Rahway, N.J., ed., Stecher, P.G., p. 958 (1968).
135. Jakinovich, W. Jr., United States Patent No. 4,220,667 (1980).
136. Bueckenhueskes, H., and Gierschner, K., *Indust. Obst und Gemusever.* 70, 3 (1985).
137. Taylor, C., *Everything You Always Wanted to Know About Pool Care*, CP Taylor Publications, Box 5866, Arlington, Texas 76011 (1974).
138. Leddy, J.J., Jones, I.C., Lowry, B.S., Spillers, F.W., Wing, R.E., and Binger, C.D., *Encyclopedia of Chemical Technology 1*, ed. Martin Grayson, John Wiley and Sons, New York, p. 831 (1978).
139. Khimicheskaya, V.D., Smirnova, E.S., Pivovarov, A.A., Steba, V.K., *Khimicheskaya Tekhnologiya 4*, 7 (1985).

CHAPTER
NINE

Induced Cake Formation

SCOPE

Induced or intentional cake formation has been the subject of many scientific articles and books. Terms such as prilling, pilling, spray drying, agglomerating, compacting, sintering, cementing, welding, briquetting, gluing, gelling, and flocculating come to mind when induced cake formation is considered. Usually when induced caking is encountered, it is a desired property while spontaneous caking is usually an undesired property. Induced caking can be classified into the same categories as spontaneous caking but caking that can cause all manner of troubles in spontaneous cakes is often much too fragile to be of use in products where induced caking is desired.

Induced caking is probably not an invention of mankind. Many animals, birds and insects utilize the technique in building their den, lair, nest, or tunnel. Mankind probably emulated the creatures around him to construct his first pots, bricks, and huts. Once the concept was presented, mankind explored and improved the methods to prepared ever better products. Science, art, and skill are required to manufacture the caked items of modern societies.

Induced caking requires greater mathematical and physical testing than spontaneous unwanted caking. Usually when a product is intentionally forced to form lumps, cakes, briquettes, containers, and so forth there is an end use in mind. This requires certain precisely specified characteristics of the product that can be a long laundry list of required properties. These properties may range from strength, color, friability, bulk density, and bulk tensile strength, to precise dimensions. These are the types of properties for which the American Society for Testing Materials (A.S.T.M.) has published thousands of pages of test procedures and specifications. Theory as well as directions for applying the tests are detailed. The reader is referred to A.S.T.M. and similar references for information and instructions for performing these tests. Test for caking properties are also presented in the A.S.T.M. publications.

AGGLOMERATION

Agglomeration differs from prilling and spray drying (to be discussed in the following) in several important ways. In prilling a liquid phase may or may not be intentionally vaporized from the hot particles falling in the prilling tower but no heat is added to the tower. Indeed, heat is being extracted from the product in the tower as rapidly as is practical. In spray drying a liquid phase, usually water, is always being vaporized from the particles falling in the spray tower and heat must be added to dry the product. In agglomeration processes heat may or may not be added depending on the product but as little heat is used as can be accomplished while as little liquid phase is added as can be tolerated while preparing a consistent agglomerate with uniform properties mechanical energy is frequently added.

The agglomeration process is very old having been employed in the ores and minerals sciences for many years. In the agglomeration process a liquid phase is used to "glue" one or more powders into granules of just about any size and bulk density desired. The process is usually carried out by spraying the liquid on to a rolling bed of powders. The beds may or may not be heated and may be contained in anything from rotating drums to fluidized beds. Usually no molds, dies, or extruders are used in agglomeration as they are in briquetting, pilling, and pelletizing.

M. J. Dolan discussed an equation to approximate detergent agglomeration as presented in *Chemical Engineers' Handbook* [140] [142]. "Many types of agglomeration can be mathematically modeled. An equation is often used to calculate liquid requirements for agglomeration by assuming that liquid binder replaces the air occupying the space between agglomerated particles:

$$X = \frac{1}{1 + (1 - v_f)p_s/v_f p_l} \qquad [24]$$

X = Weight fraction liquid in the finished agglomerate
v_f = Porosity before agglomeration (void fraction)
p_s = True particle density
p_l = Density of the liquid

If this standard equation is applied to detergent agglomeration, however, the results are not particularly accurate. Generally more liquid is needed to accomplish agglomeration than is predicted by the equation. A possible explanation for this variance is the system dynamics, which encompasses simultaneous hydrate formation, semisolid silicate deposition, and agglomeration. The fate of water and the physical state of the formulation components is constantly changing. Equilibrium is not established until late in the process. At a minimum, correction factors and a recognition of system dynamics must be added to conventional models to accurately describe detergent agglomeration."

It is remarkable that the equation comes close in a detergent formulation. If the agglomerates were composed of one pure solid substance and the liquid were a pure liquid, the equation should hold very closely, provided the solid did not solvate and change the particle density. In complex mixtures of solids, as in detergents, the true solid density can at best be an average and unless the solids have approximately the same densities the weight fraction of each of the solids and their densities must also be considered. Additionally, in a detergent mix the liquids may be a solution of sodium silicate in water and the surfactant may or may not be a liquid. It is obvious that the control parameters in a process that is expected to yield tons of materials with similar properties as bulk density, rate of solution, frangibility, color, particle size distribution, packaging, and caking must be very carefully controlled and seemingly minor variations in raw materials, order of addition, sojourn time, temperature, rate of hydration, solubility, and liquid properties could cause major variations in the specifications of a product.

The advantages of the simple equations is that it offers a starting point to initiate laboratory and pilot plant test. If too much liquid is added to a mixture to be agglomerated, the mixture becomes a soup. If too little liquid is add to the solids, no agglomeration may occur but the process can be so sensitive that the addition of very small quantities of liquid may push the system from non-adhering solids to good agglomerates, while the addition of a slightly greater quantity may produce a product that is certain to cake if transported before equilibrium has been established.

Some formulations depend exclusively on admixing solids to give a powder with the desired properties. Although it is possible in some cases to handle mixed powders, these powders usually segregate during handling and transportation. In many cases this leads to undesirable results in the final application of the powders. The larger particles tend to migrate to the top of the container while the bottom of the container may consist of fines of one of the components of the mixture.

Orr published an intriguing article dealing with dry solids chromatography. The method will be useful not only to separate solids but can be related to the segregation of mixtures when the solid mixtures are vibrated as during transportation [143]. In the experiments a column was fill with loosely fitting glass spheres. Mixed powders were allowed to pass through the column by vibrating the column. It was learned that large particles passed through more slowly than smaller particles. Long acicular particles passed more slowly than irregular shaped particles or shorter acicular particles. Interestingly, spherical particles passed more slowly than irregular particles of approximately the same size. Perhaps this was a result of the spherical particles being able to pack into voids with more contact area than irregular particles. This technique should work well in the study of mixed powders used in formulations that are not spray dried or agglomerated in other ways. Usually, both spray drying and agglomeration lead to products that not only have more aesthetic appeal but are more uniform in the quality control parameters consistent with good manufacturing practices.

There are a wide variety of reasons why agglomeration may be used in a manufacturing process. In processing ores there is often a preferred physical form of the burden entering a furnace. This may range from a porosity to gas formation in a furnace to the intimate mixing of reactants in a manner that will not allow them to segregate in a furnace before they have had time to react. Handling a burden is usually more easily accomplished if the burden is more uniform and the problems associated with excess dusting can usually be handled more readily in an agglomeration process than in other parts of the process. It is not difficult to understand why spray drying can be less difficult to control than agglomeration, particularly when complex mixtures of solids and liquids are desired in a tightly specified physical and chemical form, but capital cost of equipment and energy utilization may override all other factors when installing a new process.

Despite all of the positive attributes of agglomeration it is not without some problems of its own, particularly when complex mixtures are being agglomerated. Even the best of the agglomerated mixtures seldom match the uniformity and general appearance of spray drying. Not all mixtures are compatible for agglomeration.

Doland's conclusions are as follows. "Significant development in agglomeration technology has occurred in the last decade, at an accelerating rate. On one hand, older processes and equipment have been modified and optimized; on the other hand totally new systems have been introduced and commercialized. Installed capacity is at an all-time high and nearly 35 domestic agglomeration units are in operation (in the detergents industry). This compares with approximately 30 spray tower facilities in the U.S."

"Agglomeration offers the advantages of low energy consumption, low capital requirements per unit of capacity, low processing cost, and minimal waste. It provides aesthetically attractive products and affords the opportunity for new concentrated product forms. The hardware has reached a high stage of development. Many choices in equipment are available and a systematic understanding is evolving which will further expand the range of products suitable for agglomeration processing. The challenge for the detergent industry is to take maximum advantage of this technology."

To save space on super market shelves, the need for denser products continues in the detergent industry. More and more manufactures are turning to smaller, more densely packed products to save space on the super market shelves. This has always been an option, but in the past the customers preferred the larger boxes. Today's market is grossly changed with the demise of phosphates in detergents. Liquids have a much larger market share and the old familiar boxes have very little to do with the contents of the box. Only the box remains and though it looks as it did for years the working ingredients are not.

Strength is one of the primary considerations of agglomerated products. It has been shown by Adams that particles do not usually fail by instantaneous degradation on shear planes but that particles usually fail by a propagation of Griffith cracks (not the author) [144]. Griffith cracks are considered in greater detail in works considering tensile strength and reinforcements of material science. Because cracks are responsible for particle fracture, this can only mean that the more uniform agglomerated particles are formed the stronger the resulting particles will be to impact. The fewer flaws as cracks the less inclined the particles will become toward fracture. It should also at once be obvious that if any of the components in an agglomerated product undergo a phase transition in the ambient temperature, the particles are subjected to that the products will form countless cracks even if the constituent is a minor fraction of the agglomerated formulation.

A process named instant agglomeration was introduced anonymously [145]. The process is spray mixing. An atomized liquid is mixed with powders suspended in a turbulent air stream. It is claimed that most of the disadvantages of agglomeration and spray drying can be overcome.

Listed among the advantages are more uniform particle size distribution and simpler less expensive manufacturing plants. Although detergents were the products most discussed, it was mentioned that many varieties of products could be processed in the same equipment.

As phosphates have been discarded from the solid detergent, formulations of the future are likely to encounter three problems. The pH of the products may be high enough to cause some skin irritation. The solubility and low water holding capacity of the solids are likely to create caking problems. The very large number of components used in the emerging detergents, in an attempt to sustain past performance, is certain to cause control problems both in raw material supplies and formulations. These new solid products should agglomerate well because of the larger number of powders and liquids used in the formulations.

PRILLING

Prilling is a process in which a melt of a metal or a salt is poured or sprayed into the air of a tower. The liquid is allowed to fall a sufficient distance for the liquid to solidify before it encounters the bottom of the tower. As the liquid falls, the surface tension of the melt causes the droplets to become approximately spherical but the quality of the spheres at the bottom of the fall will depend upon a large number of factors. These factors will range from droplet size to heat transfer and freezing temperature. If the droplets have not had sufficient time to solidify during the free fall, they usually splatter or stick to form cakes upon impact. With products as ammonium nitrate or urea, a tower may be well over one hundred feet high. If a gas is liberated in the liquid that is forming a prill, the prill can become a hollow sphere. Over ammoniated anhydrous ammonium nitrate can produce "prills" that are referred to as popcorn by plant operators because of the similar appearance to popcorn.

One of the earlier applications of prilling was the shot tower in which lead shot were prepared. The invention was cleverly conceived considering the time of the invention. Required knowledge of several fundamentals was probably acquired by utilizing natural formations, for instance, water falls and cliffs, before the first towers were constructed. Some shot towers were constructed in which the lead shot fell into water at the bottom of the tower. The towers still had to be high enough to allow the shot to cool to a hot sphere but not so high that the shot could not withstand impact on to a hard surface at the bottom of the tower.

A similar approach to the rapid quenching technique is probably not practical in the preparation of water soluble salts on an industrial scale but similar techniques have been used in the laboratory to produce experimental prills when a prilling tower was not available.

Hot melts can be dropped into non-solvents provided the liquid is not flammable or toxic under the conditions of use. Carbon tetrachloride has been used extensively in the past. Under no circumstances should sodium, potassium, or similar metals be prilled while attempting to use carbon tetrachloride as a chilling liquid. An intense explosion is very likely to occur!

The making of experimental prills can be a problem when only laboratory quantities of materials are desired. Bitter experiences have caused many manufacturing plant personnel to be unwilling in cooperating in the use plant equipment for small experimental batches. Two approaches have been used to overcome these problems. If prills are made very small, they can usually be solidified in a drop of two or three floors in a pilot plant. If a large tower is available, a small tower can be constructed along side of the major tower with equipment independent of the production tower. Research dedicated towers of this kind can challenge production towers in the quality of prills produced. Production is not curtailed while using the "piggy back" tower and the danger of contaminating the production product is eliminated.

Microprills are worthy of additional mention. If the melt from which prills are to be prepared is atomized, then the melt will solidify during a much shorter drop. Although the resulting prills have much more surface area than normal prills and they do not usually flow as easily as larger prills, they have advantages of being similar to powders in many ways.

Prilling is not truly an induced caking process. It is usually considered in engineering works as particle enlargement or some similar title. It is possible under some conditions to prill a melt system that contains unmelted solids but spray nozzle plugging can become a problem even with rigid process control. Prilling can, therefore, be used to improve the flow characteristics of systems which would not flow readily in their natural crystal form. This is particularly true of acicular or fibrous crystal forms. Prilling can limit the contact area of the particles when compared to most crystals of similar particle size.

The modern prilling tower may be several hundred feet tall and may contain elaborate equipment to control temperatures, droplet size, viscosities, moisture, and so forth. It is possible to prepare prills of very uniform particle sizes and required bulk densities. It is even possible to prepare hollow prills when they are desired. Unfortunately, emission controls of prilling towers can be difficult and prilling is not as popular as in the past years.

Prilled products that contain constituents that undergo phase transitions can be troublesome. Anhydrous prills, in particular, are likely to shatter and cake during transitions. The results of repeated transitions is usually swelling first followed by powdering of the prills. After several cycles some prilled products are so degraded that it is no longer recognizable that the product was ever in a prilled form.

SPRAY DRYING

Spray drying has some of the characteristics of prilling particularly if the prilled medium is not anhydrous. Spray drying is similar to agglomeration in other respects. In prilling operations which process diluted systems, the process depends upon the latent heat of the product to evaporate any excess solvent in the prilled melt. In spray drying the product is sprayed into a chamber in to which heated gasses are blown. Rather than cooling the spray to a solid, heat is added to evaporate the solvent in the particles. The heat load for spray drying can be large and in some industries more economical methods are being sought to agglomerate complex mixtures. The exhaust gases from spray towers can contribute to air emission problems if they are not handled properly.

Spray drying is often employed when several ingredients are formulated into a product and it is desired that the ingredients do not segregate before the product is used. Another advantage of spray drying comes from bulk density control that is often very difficult to obtain by other processes. Yet another advantage in many systems is that the product is likely to be spherical. The bulk density of a product can be very important when it is desired that a specified weight of product fill a box or other container to a desired level. On most packaged goods both weight and volume must be controlled if automatic packaging equipment is to satisfy customers. No one is pleased with a box that is half filled when first opened even if it does contain the specified weight of free flowing product. If the bulk density is out of control, the denser product may well cake in the container even though the less dense powders remain free flowing.

Masters has published numerous papers which cover most aspects of spray drying in detail [146]. As fuel costs continue to rise, some strong advocates of agglomeration become even more conspicuous, particularly in the area of detergent manufacture [147]. The article by Dolan shows electron micrographs of a sprayed dried, admixed and an agglomerated product [140]. All three products are commercial detergent formulations and all products have certain advantages. Both spray dried and agglomerated products control particular segregation in the box and both can be used to control bulk density more easily than admixed products.

Microencapsulation can be accomplished in a spray tower and in some cases caking problems could be solved in this way. One advantage is that most products that are currently spray dried could probably be encapsulated for little additional cost. The technique should have much unused potential.

MICROENCAPSULATION

In recent years the techniques of encapsulation become more and more sophisticated to the degree that many pharmaceuticals are encapsulated in time release coatings while other drugs are designed to pass through the acidic stomach while dissolving in the higher pH of the intestines. There is an excellent review of this area by R.E. Sparks in the *Encyclopedia Of Chemical Technology* [156]. It is pointed out that carbonless carbon paper is still the major use of the technique where gelatin and gum arabic are allowed to coacervate in the presence of ink or other system to be encapsulated. It is reported that in 1981 the market is 500,000 tons per year for carbonless business forms. This is surely the weight of the forms rather than the encapsulated ink but even so the market is large.

Lenk and Thies studied the coacervation of gelatin and polyphosphates and demonstrated that the condensed phosphates were very effective in interacting with gelatin [148]. Because the "hexametaphosphate" of industry is a mixture of polyphosphate chains that are strong electrolytes, the polyphosphates they used in their study should have an average of about 10 negative charges per molecule. The coacervates of gum arabic and gelatin contain about equal weights of both reactants but as little as 10% or less of the weight of gum arabic is required to separate a gelatin coacervate with the polyphosphate used in this study. The polyphosphates have been used for many years to tan leather by precipitating the gelatin in the pores of an animal's hide. The gelatin-polyphosphate coacervate is highly refractory and helps to waterproof the leather [149].

The polyphosphate coacervates with gelatin and other proteins will surely become important in the future for carriers of special proteins in biological systems. It has long been practiced in bone scans that either polyphosphates or phosphonates are used as carriers to transport technetium to the site of a tumor where it collects and can be monitored and detected. There is no reason why the polyphosphates cannot be used in the future to transport special proteins or other therapeutics to the site of a tumor or other organ where the rates of metabolism is increased and phosphatase activity is very great.

TABLETING

Tableting or pilling has been practiced for years as a form of induced caking designed primarily to furnish powdered medicines and other ingredients in forms that were precisely measured. In this way a supply of predetermined dose can be prepared while delivering the product in a form that is convenient to store and use. Products as diverse as insecticides and special detergents to aspirin and candies come to mind.

Most powdered products can be made into a tablet form and even substances as diverse as acids and bases can be contained in a single dry tablet that effervescence when placed in a glass of water. Mixtures of monosodium orthophosphate and sodium bicarbonate or citric acid and sodium bicarbonate work very satisfactorily in this respect. Some powders resist most efforts to press them into tablets and a secondary substance must be added as a binder to cause to product to form cakes strong enough to withstand handling and transporting. Binders can also add hygroscopic problems that were not a part of the initial system and care must be exercised in choosing a binder.

Typical binders have been chosen from food type products for most applications. Starches, sugars, gums, glues, and gelatin have been popular. This is a result of the wide use of tablets in the pharmaceutical industry where food grade materials are required. One of the major problems encountered in preparing tablets of a specified weight is rapidly filling a die with the proper quantity of the powder to be pressed into the pellet. This means that it is imperative that the powders are delivered to the tableting machine in a free flowing uncaked condition. To avoid some of the problems many substances are predensified as very large tablets that are then ultimately powdered before being fed to the tableting machine. Aspirin was often sold in the past by suppliers as tablets weighing more than one pound each. This technology has mostly been superseded by roll compaction and the compacted mixture of aspirin and starch is fed to the tableting machine. The major problem remaining with the manufacture of aspirin tablets is called filming. Most tablets are embossed with either emblems or letters and filming causes the impressions to be deformed or otherwise unpleasing.

CASTING

As mentioned, much of the industrial and institutional (I and I) segment of the detergent industry cast the detergents to help them to perform in automatic equipment. Many of the chemicals used in swimming pools may also be sold as large cast cakes. A few of the patents are worthy of review.

Miret Plaja employed an hydraulic press to prepare cakes of trichloro isocyanuric acid. Pressures of approximately 110 kg/cm^2 yielded strong cakes, which are claimed to dissolve slowly [150]. Bridges claims that calcium hypochlorite can be agglomerated with a turbine agglomerator. The calcium hypochlorite was agglomerated with water and then dried to about 8.1% water in the final product. The product is also claimed to dissolve more rapidly than anhydrous calcium hypochlorite [151]. Granular calcium hypochlorite dihydrate is claimed to be non-caking if it is sprayed with an aqueous solution of alkali or alkaline earth chlorides or nitrates. The product is claimed to be coated when dried [152].

COMPACTION

Compaction takes on a whole new dimension when compared with the products and industries have been considered. Most areas of civil engineering are gigantic when compared to processes. There are often times that highways, runways for airports, railroads, bridges, dams, and buildings are required in locations where the subsoil is not of the proper consistency to support the desired structure. Even the compaction of tailings in ponds used in the mining industry are a part of the general science of compaction where induces caking is desired.

Most compaction is induced by mechanical and chemical treatments but electroendosmosis has been successfully applied in a wide variety of construction projects as well as in tailing ponds [153]. Electroendosmosis occurs when electrodes are placed in a medium capable of behaving as a membrane. Clays and silts are usually good examples. An electrical potential is placed across the electrodes and the continuous medium, water in most cases, will migrate toward the electrode with a charge the same sign as the sign of the charge on the medium acting as a membrane. In the mineral world the charge on the membrane is usually negative and water will, therefore, migrate toward the negative electrode. As more and more water migrates from the clay or other solids, the solids become more and more compacted. This technique could become very important in years to come. In the attempts to remove sulfur dioxide from the stack gases of power companies, other problems are being created. One of the typical methods employed is to scrub the stack gas with lime slurry. This creates millions of tons of slightly soluble calcium sulfite. The only economical way known to dispose of the white slurry is in tailing ponds. Already some of the largest tailing ponds ever constructed are filling with a useless white colloid. Too many attempts to solve one problem are creating others as bad or worse.

Chemical compaction has mostly centered about quick lime, CaO, as the agent of choice to stabilize soil. Some of the published work attributes the action of lime in aiding in compaction to the basic chemical attack of lime on silicates of clays [154] [155]. This may well be true but colloid science cannot be ignored. It has long been recognized that calcium ions are good flocculating agents for soils and clays. It was mentioned above that wet soils will normally exhibit a negative charge. This is because the frame work of most clays, silts, and so forth, are made from very large anions, silicates for example. The anions cannot easily migrate away from the particles since they are the framework of the particles. Some of the cations associated with the clay are free to migrate away from the particles leaving them negatively charged particularly if the cations are singly charged. If the negative charge on the particles is increased by adsorbing hydroxide, polyphosphate, or metasilicate anions, the particles repel each other and the system is deflocculated. Multi-charged cations behave in exactly the opposite way when mixed with negatively charged colloidal clay particles. They cause the negatively charged particles to attract each other and as mechanical pressure is applied to the flocculated particles the soil is stabilized and compacted.

The previous discussion is based upon the concept of the Helmholtz double layer. When a particle will migrate toward a positive electrode, the particle is said to have a negative charge. The negative charge is within or behind the shear plane of the continuous phase and the particle. If the particle is placed in an electrical field, then the particle is attracted toward the positive electrode and if the medium is dilute enough, the particle will migrate toward the positive electrode dragging the negative charge on the particle. But the theory explains that there is a swarm of positive ions on the medium side of the shear plane. In this case the positive ions are free to move about like bees about a hive. Two factors influence the behavior of the cations, concentration, and charge density. The negatively charged particles normally repel each other and are, therefore, deflocculated. If a sufficient number of multi-charged cations are dissolved in the continuous phase, they can overcome the repulsion of the particles and cause them to adhere to each other. As they contact and interact with each other, the compaction and strength becomes greater allowing the treated soil to support more load than the native soil.

This is the probable mechanism of compaction rather than an attack on the silicates per se. Most clay systems are reasonably resistant to attack by alkalies at lower temperatures. A chemical attack on the silicates by calcium oxide implies that there is a depolymerization of the molecular structure of the silicate rather than a simple ion exchange of calcium ions for hydrogen ions and alkali metal ions. There is small doubt that ion exchange occurs and it can be demonstrated that the charge on a particle can be reversed from negative to positive by treating the particle with multi-charged cations. This is to say that the negative colloid can first be flocculated by the addition of cations and then deflocculated as more cations are added and particles that were first attracted toward a cathode are then attracted toward an anode.

A very interesting and informative investigation that can be applied to systems that are intended to be compacted is the determination of the zero fluidity point of the soil as a function of soil concentration and conditions. In order to make the measurements, all of the solids must be small enough to pass through a capillary tube. Zero fluidity is that concentration at which the solids concentration in a colloidal suspension are great enough for the solids to flow as a unit rather than as independent particles.

If one plots the *volume percent solids* versus the reciprocal of the flow time obtained with an Ostwald viscometer for a number of colloidal suspensions of different concentrations, a straight line is obtained. An extrapolation of the straight line to the concentration axis yields the zero fluidity point. The volume concentration can be easily converted to weight percent solids from a knowledge of specific gravity of the solid particles. The zero fluidity point will be a function of the concentration of either lime dissolved in the water used to make up the solutions or of deflocculating agents used to treat the solids. A deflocculated system will have a higher zero fluidity point while a flocculated system will have a lower zero fluidity point. True zero fluidity is infinite viscosity but the extrapolated system merely measures the concentration at which the suspension will no longer run through the capillary of the viscometer because it would plug the capillary of the viscometer.

SUMMARY

Converting powders and mixtures of solids to granules, prills, tablets, and so forth has many practical applications in helping to keep systems of this type free flowing. For many years the detergent industry and the ore and metals industries have employed these techniques in order to improve their processes and products. Recent trends in the detergent industry have been toward agglomeration while in the past spray drying detergents was considered the preferred way.

Prilling has been used as a method of preparing urea, ammonium nitrate, and other fertilizers with low melting points or very high solubilities to prepare solids that handle easily and have bulk densities in the range desired for the particular products. In the feed industries where bulky, fibrous materials must be handled, they may either extrude or otherwise pelletize mixtures for chicken feeds or similar products. Tableting is practiced much in the pharmaceutical industry and is an art form that must be tailored to each product but some general rules do exist.

Compaction is an area unto itself. In this case roadbeds, dam footing, runways, and similar construction sites are prepared to handle loads by inducing a cake formation. Usually both mechanical compaction and chemical stabilization are used with lime being a preferred stabilizer.

REFERENCES

140. Dolan, M.J., The Soap and Detergent Association 1987 Annual Convention, Boca Raton, Florida; *Soap/Cosmetics/Chemical Specialties*, March, 33 (1987); ibid., *April*, 45 (1987).
141. Berg, P.J. and Hallie, G., *Proc. Fertilizer Soc. 59*, (1960).
142. Perry, R.H. and Chilton, C.H., *Chemical Engineers' Handbook Edition 5*, pp. 8–57 to 65, McGraw-Hill Book Company, New York, N.Y., 148 (1973).
143. Orr, C., *Powder Tech. 50*, 217 (1987).
144. Adams, M.J., *J. Powder Bulk Solids Tech. 9*, 15 (1985).
145. Anonymous, *Process Engineering 68*, 21 (1987).
146. Masters, K., *Spray Drying Handbook, Ed. 4*, Halested Press, New York, N.Y. (1985).
147. Kuti, S.A., *J. Am. Oil Chem. Soc. 55*, 141 (1978).
148. Lenk, T. and Thies, C., *Coulomic Interactions in Macromolecular Systems*, Ed. Eisenberg, A. and Bailey, F.E., American Chemical Society, p. 240 (1986).
149. Van Wazer, J.R., *Phosphorus and Its Compounds*, Interscience Publishers Inc., New York, p. 467 (1958).
150. Miret Plaja, A., Spanish Patent No. 550348 A1 (September 1987).
151. Bridges, W.G., US Patent No. 85-717983 (March 1985).
152. Faust, J.P., US Patent No. 3953354 (April, 1976).
153. Staff, *Chem. Eng. News 62*, January 30, p. 23 (1984).
154. Boynton, R.S., *The Chemistry and Technology of Lime and Limestone*, p. 196, John Wiley and Sons, New York (1967).

155. Boynton, R.S., *Kirk-Othmer Encyclopedia Of Chemical Technology*, 14, p. 377, Ed., Martin Grayson, John Wiley and Sons, Inc., New York, N.Y. (1981).
156. Sparks, R.E., *Kirk-Othmer Encyclopedia Chemical Technology* 15, p. 470, Ed., Martin Grayson, John Wiley and Sons, Inc., New York, N.Y. (1981).

CHAPTER

TEN

Overview and Outlook

The area of solids handling will continue to improve and many new fields will become a party to the improvements. Perhaps one of the changes that should be expected is powders behaving more like liquids. In most industrial environments liquids are much more easily handled. Several examples of solids behaving much like liquids have been cited. The prilled ammonium nitrate that had been treated with MgO became so fluid that it could not be contained in the conventional spreaders because it ran through the cracks in the machinery. Ground limestone that had been treated with oleic acid became so fluid as to be uncontrollable in equipment that had been used for years to pipe the calcium carbonate from storage bins to mixing rooms. It has been shown that powders can be easily charged with glow discharges to make them at least measurable by electrostatic sensors. This technique could be used to assist in the propelling of powders in an inert gas stream in much the same way powders are repelled by a charge when being weighted on balance during dry cold weather.

At the risk of being too futuristic, several of the areas that should be responsible for improvements during the next two decades and some of the problems that will probably be encountered are listed below. Following the list, selected items will be discussed in some more detail to complete the study of the caking of solids.

1. The new fields of microscopy, such as tunneling microscopes and force field microscopes, will yield much more information about the conditions of surfaces than have been available until today. Even light microscopes continue to improve and their usefulness cannot be over emphasized. The optical microscope is the very first equipment that should be utilized in any caking problem. There is nothing that supplants seeing a work and knowing what is being dealt with, first

hand. Even if you have help to do the laboratory work be certain to *look yourself* at what the samples look like when viewed under magnification.

2. Greater understanding of solubility will add to the predictive powers of what the influences of changes should be. Our fundamental understanding of this important area is pathetic. It is at the very heart of much of the science called chemistry. Much time is wasted searching for a solvent. Phosphates have been discussed throughout this book. There is but one known solvent from alkali metal and alkaline earth polyphosphate salts; water. Is this the only possible solvent? If so, why? It is at least needed to be understood that there is no reason to continue to search if there is no possible solvent.

3. Hydrate formation and decomposition needs additional work to better understand and control these systems. Here again there is no method of even predicting whether or not an anhydrous salt will form a crystalline hydrate or other solvate. A knowledge of methods to predict the behavior of salts could do much to help in the formulation of new products. Most of the time results are learned by mixing hundreds of samples and then testing them by subjecting them to *jungle room* conditions for days or weeks. This is a labor of ignorance and if the fundamental knowledge to prevent this waste of time and effort is ever to be obtained, it must be initiated by the industries that will gain from the work. It is a type of work that is too basic for most academics who prefer to work at the leading edge of the art forms of science, preferring to shun the stigma of applied research.

4. Ways to predict when double salts and solid solutions can be expected in melt systems and in aqueous systems are needed. A method of predicting *a priori* whether or not the tie lines in the phase diagram will or will not slope in the direction to decrease the hygroscopicity and caking of a solid solution.

We have passed these areas by as though we had gleaned all of the knowledge we could ever expect to learn rather than approaching the problems as necessary building blocks for any real understanding of our systems.

5. Obviously it is known that the shape of particles influence both their flow and caking characteristics but surprises are continuously encountered and a better understanding of the factors involved with flow and caking is needed but some interesting work has been done in this area and more is likely to be done. This has occurred partly because of continued interest in materials science. It is absolutely necessary that we find materials with new properties if we are to continue to build on the systems that serve us in modern societies.

Although some groups seem to believe that we have already explored this area to exhaustion, we have hardly begun. The possibilities are almost limitless. Silicate and phosphate chemistries are still new and both chemistries are certain to yield safe new products in the future. Silicate chemistry in particular is in need of unifying work similar to the research done with phosphates.

6. Encapsulation of powders is in its infancy and more work is needed in the process of spray drying where it is already used to some extent and in agglomeration where it should do much to protect a product from environmental changes and other constituents of formulations where undesirable interactions can be encountered.

7. Pipeline transportation of solids over distances as great as coal slurries, gas, and oil should become common as highways become more congested and new method of transporting goods are required. Electrostatics are likely to play a role in this area as has been demonstrated in some practical test with bulk measurements.

8. Today's containers are far from satisfactory. Many groups are continuously badgering the solids industry to find better and more acceptable ways to package their products. The search for new methods is certain to spawn new and better ways to handle the industry. Much progress has been made in handling solids in bulk and reducing warehouse inventories during the last two decades. As communications continue to improve and the use of computers and statistics come into wider use less time will be wasted on products that have grown difficult to use merely because they remained in storage too long. In years gone by it was necessary to keep a backlog of product in order to be capable of filling orders as they were received. Today an industry is much more likely to manufacture bulk products on demand. This cuts inventory and insures fresher goods to the customer and vastly decreases the waste disposal problems of the past when the bags that had contained the products were either burned or simply discarded to any fate that might happen.

9. The improvement in equipment is continuous. The state of a product will be much more easily monitored and corrective measures may be made while the product is in transit either in the plant or in commerce. Equipment to aid in the correction of misused or stored products will also be improved. The need to recycle goods that do not meet the customers satisfaction will require that the plants of the future be built with the ability to recycle built into the process. This will minimize cost and time in rehandling goods that cannot be disposed of because of state and national regulations.

10. Personnel training will become much more a requirement of the industry hiring the workers and much less the responsibility of society. It will be expected that a new employee will spend from one to three years in the industry classrooms before actively engaging in the manufacture of goods or handling of company business. This will at first seem expensive but will in time reward both the company and the employee. Schools and universities can furnish the basic education but *training* in science, engineering, manufacturing, marketing, and management must become a *before* job training rather than an on-site training that is usually a careless, unorganized exercise at best. Each industry has its own unique requirements.

The training should also include a course on management that emphasizes the science and technology that they are managing. It is rare in a chemical company to have courses offered in a branch of chemistry. Courses are offered in all other types of areas from computers to economics and safety but too seldom are courses taught in chemistry, physics, or mathematics. Most industries will support their workers in taking courses at universities but it is not the course work that is unique to the industry and can only be learned from the particular industry.

11. Items as simple in concept as improved methods of measuring mass or weight in bulk quantities will do much to improve products. Measurements of solids in slurries are particularly difficult to handle precisely despite the fact that there is equipment that works well for some systems but not for others. Gamma ray absorption is one of the methods showing promise but still is not as reliable as desired in some cases.

12. Statistical and other mathematical approaches shall do much in the next two decades to improve both products and processes. The advent of hand held calculators and computers have changed the complexion of this science from an area that few could afford the laborious calculations to one in which a few minute's time can accomplish the work of days in recent passed years. The statistical approach to industrial problems has been very helpful but the concepts are beginning to be dated, and it is surely time for new approaches to these areas. It is time for a completely new concept that is not based on probability as its under lying logic.

13. New states of matter will do much to yield better new products and to improve old ones. It is necessary, however, that a renewed interest be gained in the search for new forms of new matter with startling new properties. One problem that has been encountered in recent years is an almost dislike for new states of matter with new

properties by industrialists and environmentalists alike. The industrialists are very cautious in spending on new products and properties. Only products that are so well known as to be completely risk free can be considered without extensive expenditures. These products are not accepted because they are safe but because they have become so familiar as to be accepted as safe. It must be admitted that management's reluctance to spend for new products is based upon some unfortunate experiences.

14. In transit treatments and shelf life, rejuvenations should become a concept of the future. Today, in some cases, desiccants such as silica gel are put in products as separately segmented inclusions to increase the shelf life and crispness of the product. The results have been very good and products that would surely perish in a few weeks unprotected can remain useful for years when properly sealed. There must be extensions of the general concept of post manufacturing protection to bulk shipments as large as tankers or rail cars. In transit ammoniation of products that may be unstable could be accomplished in a number of ways, which are no more intrusive than the application of anhydrous ammonia to farm land. More activity must be addressed to pairing products that will have the ability to protect each other. Much more attention must be given to the fate of a product if it becomes involved in a mishap during transit. Is it safe for the public? Will it cake if it is exposed to the atmosphere? Have the consequences of shipping a product in bulk been considered? This will become more and more a requirement of the future, not only from the viewpoint of time and cost but also in the elimination of waste. These changes will reform total industries and it is likely to occur very rapidly.

15. Regulations are an area that must be reconsidered during the next two decades if any progress is to be made with powders or most other products of American industry. The Japanese, Russians, and Chinese are continuing to work in the area of new material while the Americans and Europeans seem to be content to make do with the products developed thirty years ago. Regulations and fines are so severe as to cause even the bravest of industrialists to ponder the advantages and liabilities associated with a new product, even if it is an old product that no longer cakes because of minor changes in the product or process. This barrier is probably greater than the scientific barriers in many product lines. It is not yet certain what are the ultimate consequences of the ever increasing number of American companies that are controlled by foreign interest. Perhaps the necessary research shall be done abroad and imported into the United States after the development work is completed.

FUTURE IMPROVEMENTS

Future improvements that are based purely on technology in the solids industry will depend on a greater knowledge of the systems used as products. Research devoted to understand the general science of caking will be required and it must be done by the industries gaining from the knowledge. Simply, this is because nobody else will be motivated to gain the knowledge in an area that is often considered to be scientifically mundane or boring. This is difficult research to support in an industrial setting because it is difficult to tie it into an accountant's spreadsheet in such a way that the profits are definitely earmarked as resulting from a study that yielded only information. Information is all any research project ever yields. It is a learning process and it is the judicious use of the new knowledge by management, engineering, and manufacturing personnel that truly produces the new product. It is useful new products that create revenue, but the complete chain of command is required to launch a new product and the new product depends on all parts of the enterprise to emerge successfully.

NEEDED LABORATORY WORK

A study of seed crystals and nucleation phenomena in general deserves much more work that is sufficiently understood to be expressed in physical terms and concepts employed by the work-a-day laboratory technicians. These are the personnel who must ultimately translate the information into workable materials and hardware. It has been seen that the pyroelectric crystals and the ferroelectric crystals have an electric field surrounding them that can be observed by allowing MgO dust to settle on to the electric field. Do these fields influence the growth of crystals from melts or is the influence of a seed crystal merely a templet on which new crystals can grow? The influence of seed crystals upon the growth of sodium Kurrol's salt of phosphate is an interesting example. Without the proper seeding sodium Kurrol's salt does not exist. Only certain seeds will cause the system to crystalize as the long chain polyphosphate with crystal properties similar to asbestos. The seed crystals on the phosphate melts develop a region around the crystals that seem to have the structure of an electric field similar to the electric field about active pyroelectric crystals. It will be remembered that seeding of a system can set off transitions that can lead to severe caking of products.

Much new information is needed to understand more about the fundamentals of agglomeration. Despite all of the published literature on the subject, a current project has degenerated to a trial and error process that is expensive and inefficient. There is little doubt that a method will be found that will meet the requirements but several man months shall be spent that could better be spent in other areas if the agglomeration of products was better understood. The same comments could be made about the other granulation areas.

The surfaces of solids are definitely within the province of colloid science but not nearly so much has been done with dry solids as has been accomplished with colloids contained in a liquid medium. There is probably a behavior in solid powders that is not too unlike the Hofmeister series when applied to aqueous colloids. In the flocculation of colloids, the Hofmeister series of ions is arranged in an order where multivalent ions are much more active than univalent ions. In one sense the flow conditioners are similar to deflocculating agents used in suspensions and slurries but no organized set of rules have been formulated to guide the laboratory worker as to the type of substance that should be considered with a product of defined properties. The first attempts usually include tricalcium orthophosphate, fumed silica, powdered graphite, or magnesium oxide. Fortunately, some interesting work is in progress in this area and greater understanding is certain to be forthcoming.

Friction of flow is getting some well-deserved attention on how solids behave in bins and in conveying equipment. Also the concepts of lubricity on solids are being considered and this will add to the knowledge of flow conditioners as well. Along with the other electrical properties of crystalline solids is the dielectric constant of solids. This not only plays a part in the caking of solids but the higher the dielectric constant the greater the charge the crystals are likely to have. On the other hand, the low dielectric solids as plastic molding powers are likely to have low energy surfaces that are very poor conductors, and as a consequence, are likely to build up charges during air conveying and other means of moving in equipment. Much more is needed in the chemistry and physics of static charges both in powder technology and in the fabrics and carpeting industries.

Static charges are particularly worrisome with much of the new electronics equipment that depend upon chips that are easily destroyed by bolts of static electricity. The charge can be created simply by a worker walking across a carpet before activating the equipment. This is not directly related to caking but it is a part of the problems occurring in the transporting of solids and the unloading of equipment that is caked because of static charges.

Quality improvement teams can be very useful in controlling caking problems of products. A primary usefulness of a team can be to point out existing and potential caking problems and to find the customer's response to caking problems in general. The impetus for the QIT is directed toward the expected problems to come as changes are made away from old known substances to new products with more acceptable environmental and safety properties than the old products had. Many opportunities should exist for the application of sound techniques in the chemistry and physics of solids. Controlling the caking and flow properties of the many new products of the future will require new insight of the fundamentals of the physical world.

A QIT composed of up to twelve members ranging from research workers to sales, manufacturing, and statisticians can be very useful. It will soon become obvious that if known caking problems are to be solved in a manner that can be implemented it will require the input of all members and no one type is any more or less important than the others. It has already been seen that there may be many ways to satisfy a customer's needs and that a customer may be willing to live with hardships that were at first believed to be unlikely, provided the customer induced the hardships on himself.

At this stage of history there is much positive emphasis on the team concept even to the extent of attacking problems as mundane as caking. It is probably true that teams perform much better in development and manufacturing functions than in the research areas. As Phillip Wildy so apply noted in his classic discourse, *A Generation of Vipers*, a large group has never painted a picture, written a great book, written great music, invented a great new product, or had a great thought. These are more in the purview of the individual. The solving of caking problems will also be more in the domain of the individual than in the group approach but the managing of the approaches to solving the problems may well lie with the collective thought processes of the groups. It is necessary to always keep in mind that there is nothing democratic in the laws of Nature and whether a product is free flowing will depend upon the properties of the product and not the consensus of a group. But the customer may well be a group, and it is difficult at times to know what a customer will and will not accept. This reiterates the extreme need to know what the customer wants and needs.

INDEX

A

Abrasion 18
Absorption 21, 27, 35–45, 55, 59, 60, 71, 74, 110, 147, 148, 187, 226
Acetates 72
Acid Magenta 123, 191
Acrylic polymers 193
Activator (bleach) 73
Adenosine triphosphate 59
Adsorption 39, 42–45, 48, 78, 112
Agglomerate 3, 209, 213, 215
Aging 6, 27, 119
Alkyl benzene 50, 57
Allotropic forms 84
Ammonium nitrate 3, 6, 9, 12, 19, 21, 26, 36–55, 71, 84, 87, 99 101, 107, 120–123, 126, 133, 167, 185–195, 213, 221, 223
Ammonium phosphate 133, 167, 191, 192
Ammonium sulfate 50, 120, 186, 191–193
Amorphous solids 4, 39, 46, 47, 58, 59, 70, 75, 135, 161, 162
Apatite 64, 196, 197
Asbestos 122, 228
Aspect ratio 4, 18
ASTM 143, 168

B

Baked products 23
Barium titanate 124
Blasting 6, 101, 185, 186, 188, 191, 194
Bleaches 187, 201, 202
Boric acid 190
Bottle set 3, 126
Briquetting 208, 209
Brown sugar 186
Bulk density 4, 18, 27, 33, 54, 55, 88, 128, 208–210, 215, 216

C

Cab-O-Sil 133
Calcium carbonate 71, 126, 127, 191, 202, 223
Calcium chloride 21, 45, 46
Calcium phosphates 191, 196
Camphor 23, 48
Carbon dioxide 19, 22, 23, 44, 72, 74–76, 78, 197, 198, 202, 204
Carbonization 16, 74, 181
Carbowax 133
Casting 5, 218
Cement 3, 22, 28, 29, 75, 76, 87
Ceramics 1, 47
Chaos 54, 83, 113
Charcoal 129, 130
Cheese 31
Chemical caking 16, 19, 31
Chemicals 1, 24, 33, 126, 127, 141, 187, 201–203, 218
Chemisorbed 43, 44
Classes of cake formation 15, 17–20
Clay 48, 50, 75, 132, 189, 193, 218–220
   Attapulgite 132
   Bentonite 132, 133
   Kaolin 18, 50, 132, 133
   Talc 132
Coal 1, 19, 22, 29, 187, 225
Colloids 49, 50, 126, 229

Compaction 28, 217–221
Components 97–99, 103, 111, 129, 151–153, 210–213
  Acidic 23, 181, 192, 203, 216
  Basic 5, 23, 29, 137, 138, 160, 181, 192, 202, 219, 224, 226
  Low melting 24, 28, 221
  Phase rule 5, 13, 14, 61–65, 70, 81, 82, 90, 91, 94, 113
  Unstable 24, 52, 64, 65, 86, 117, 188, 196, 202, 227
  Volatile 19, 22, 23, 141, 182
Compression caking 27
Concretus chemistry 2
Conditioners 17, 122, 132, 133, 187, 200, 229
Cooling curves 95
Copper nitrate 189
Cost
  Caking 1, 5–7, 9, 17, 26, 77, 112, 142, 143, 189, 211–216, 225, 227
  Capital 185, 211, 212
  Disposal 1, 225
  Encapsulation 198, 216, 225
  Fuel 22, 157, 187, 188, 191, 194, 195, 215
  Hidden 5, 7
  Quality 1, 2, 5, 10, 15, 33, 53, 116, 140–147, 162, 163, 168, 179, 203, 211–214, 229
  Research 5–8, 10–13, 62, 77, 138–142, 169, 187, 214, 224–230
  Safety 11, 133, 142, 146, 186, 226, 230
  Time 6, 9, 137
  Transportation 1, 6, 19, 21, 29, 137, 189, 192, 210, 211, 225
Crystals
  Acicular 18, 28, 122, 132, 211, 214
  Anhydrous 3, 6, 8, 25, 38, 40, 41, 45, 46, 51–56, 60, 72, 81, 99–101, 105, 152, 188, 190, 191, 198, 213, 215, 218, 224, 227
  Bridged 3, 16, 56
  Chemical properties 9, 28, 82
  Dendritic 24
  Domains 114, 125, 163, 168
  Ferroelectric 20, 31, 71, 86, 114, 115, 121, 123–125, 135, 163, 164, 167, 168, 228
  Ferromagnetic 86, 114, 124, 125, 163, 168
  Fibers 17–19, 70, 79, 84, 113, 180
  Habit 25, 81, 200
  Hierarchy 114, 163
  Hydrates 8, 12, 34, 38, 39, 41, 44–46, 51, 52, 55–58, 66, 71, 72, 75, 76, 78, 83, 107, 111, 116, 123, 149, 150, 152, 160, 179, 183, 185, 193, 196, 199
  Impurity 13, 22, 36, 54, 141
  Lattice 25, 45, 47, 57, 67, 78, 84, 85, 108, 117, 125
  Metastable 61, 63, 67, 86, 122
  Neutral 34, 44, 47, 48, 70, 193
  Paramagnetic 86
  Perovskite 124
  Piezoelectric 15, 20, 31, 42, 71, 78, 114–118, 120, 121, 123, 135, 163–167
  Plate 18, 54, 127, 166
  Pyroelectric 15, 20, 31, 71, 114, 115, 119, 121–125, 135, 163, 164, 166–168, 228
  Seed 52–54, 121–123, 228
  Structure 20, 23–25, 39, 47, 57–59, 82–84, 99, 103, 122, 129, 130, 132, 146, 161, 218, 220, 228
  Surface 8, 17, 18, 26, 28, 29, 37, 38, 41–50, 53, 55, 56,

57, 60, 64, 68, 71, 76, 78, 83, 84, 99, 102, 107, 110, 116–118, 122, 123, 126, 127, 129–135, 139, 148, 166, 167, 188, 191, 192, 197–199, 204, 213, 214
Templet 122, 228
Transitions 3–10, 12–20, 25, 31, 38, 42, 48, 55–61, 71, 78, 81–90, 112–117, 123, 161, 163, 181, 185–191, 202, 215, 228
Curie, Pierre and Jacques 20, 120, 136
Point 86, 168
Customer 1, 5, 7, 9, 22, 27, 31, 126, 128, 137, 140, 168, 202, 205, 225, 229, 230

D

Degrees of freedom 52, 61–69, 91–99, 149
Dehydration 3, 15, 19, 25, 40, 52, 83, 181, 182
Deliquesce 35, 37, 42, 48, 183
Detergent 5, 12, 30, 34, 56, 73, 85, 95, 107–112, 129, 193, 197, 198–221
Dew point 55, 148
Dicalcium orthophosphate 24, 25, 40, 196, 197
Dielectric constant 124
DNA 18, 122
Double salts 96, 104, 111, 152, 194, 196, 224
Dyes 1, 25, 187, 189–191, 204

E

Electrets 20, 119, 135
Electric fields 115, 119, 121–125, 135, 166
Electrical caking 16, 20, 31, 115, 125, 181
Enthalpy 61, 66, 86
Entropy 61, 86
Environment 1, 2, 45, 59, 115, 149

Equipment
 Laboratory 35, 90, 120, 129, 139, 142, 145, 154, 157, 165, 223, 226
 Plant 8–15, 29, 211–218, 223, 225
 Industrial 5, 12, 26, 30, 77, 123, 128, 185–191, 199, 211–215, 225
 Electronic 13, 120, 163, 165
Eutectics 54, 101, 106, 190
Explosives 1, 185, 187, 205

F

Faraday 53, 206
Farming 5, 146
Fertilizers 1, 6, 14, 17, 30–32, 50, 81, 132–135, 187–196, 221
Flocculating 208, 219
Flow of solids 29, 77
Food 7, 14, 21, 23, 31, 50, 133, 186, 192, 195, 198–201, 205, 207, 217
Formamide 194
Formulations 13, 23, 24, 71–79, 81, 197, 210–225
Friction 18, 19, 21, 229

G

Glass 24, 44–49, 58–61, 65, 66, 85, 86, 101, 122, 134, 148, 152, 162–166, 179, 182, 203, 211, 217
Glue 3, 209
Gold 48
Grains 29, 75, 186, 188
Graphite 18, 84, 85, 118, 119, 229
Gravity 2, 3, 28, 30, 68, 78, 194, 220
Griffith cracks 212
Grinding 50, 58, 70, 71, 79, 182, 204
Gum Arabic 25, 216
Gun cotton 11

H

Heat
- Capacity 25, 26, 41, 51, 56, 58, 76, 82, 107, 156, 157, 185, 212, 213
- Of condensation 43, 55, 68, 107
- Of hydration 3, 38, 41, 45, 51, 55, 56, 67, 107, 198, 210
- Of solution 34, 45, 46, 55, 58, 60, 63, 66, 67, 107, 192, 203, 210
- Of vaporization 41, 43, 45, 46, 55, 66, 107

High energy bond 59
Hygroscopicity 3, 16, 46, 74, 97, 108, 110, 112, 193, 224

I

Invariant point 92, 104, 105, 149, 152, 154

K

KCl 39
Kinetics of cake formation 26, 27
Knots 4, 18
Kopp's law 41
Kurrol's salt 122, 228

L

Laboratory
- Personnel 10, 27, 117, 137–141, 144, 214, 226, 228
- Techniques 12, 16, 90, 123, 137, 140, 151, 162, 187, 214, 216, 230

Leather 216
Lorentz-Lorenz equation 124

M

Magnesium carbonate 132, 198, 200
Magnesium nitrate 46, 101, 102, 133, 189–191
Magnesium oxide 99–101, 126, 131–133, 156, 166, 167, 190, 229
Managers 137
Mechanical caking 4, 16–19, 31, 79, 180
Menthol 23, 48, 96
Mercury heart 134
Methylcellulose 47
Microprills 214
Microscopy 90
- Becke line 162
- Birefringent 115, 164, 165

Microwave 186
Mixed systems 58, 69, 79
Molar polarization 124
Molar refraction 124, 163
Monoammonium orthophosphate 116–120, 123, 133, 135, 165, 166
Monocalcium orthophosphate 24, 196, 197
Mylar 134

N

NaCl 35, 39, 70
Nodulize 3
Null hypothesis 146
Nylon 12, 129

O

Oleic acid 30, 50, 126, 128, 223
Ores 29, 187, 209, 211
Orthophosphates 40, 49, 116–118, 122–124, 192, 196
Oxalates 72
Oxamide 48, 110, 193–195

P

Paraffins 48
Paraformaldehyde 48
Peracid 73
Perborate 73, 197, 198, 202
Petrographic microscope 115
Pharmaceuticals 133, 142, 216
Phase

Diagrams 12, 36, 39, 53, 54, 65, 70, 81, 82, 85, 91, 92, 96, 97, 102–104, 112, 152, 154, 157, 160, 200
Rule 5, 13, 14, 61–65, 70, 81, 82, 90, 91, 94, 113, 139
Transitions 3, 5, 6, 10, 12, 15, 16, 19, 20, 25, 31, 38, 42, 48, 55, 57, 59, 61, 71, 78, 81–88, 90, 112, 116, 117, 123, 161, 163, 181, 185, 188–191, 215, 228
  Discontinuous 85, 86, 90
  Endothermic 45, 46, 61, 66, 86, 159
  Exothermic 46, 58, 61, 66, 67, 86, 107
  Hysteresis of 168
Phosphate fertilizers 192, 195, 196
Phosphorus 48, 58, 59, 66, 72, 77, 79, 80, 117, 118, 136, 192, 195, 196, 204, 206, 221
Phosphorus pentoxide 58, 59, 72
Photographs 87, 162, 163
Pigments 1, 204
Pill 3
Plastic Flow Caking 19, 59
Poling 125
Polyethylene 42, 129
Polyvinyl alcohol 47
Prill 3, 38, 99–102, 190, 191, 213, 214
Project duration 138
Pulp 18
PVC 128
Pure Systems 11, 13, 21–24, 34, 36, 52, 54, 62–69, 78, 79, 82, 93–95, 101, 106, 108, 126, 150–159, 188, 192, 200–210

Q
Quartz 120

R
Rail cars 8, 10, 22, 72, 137, 227
Reactions 15, 16, 19, 23, 24, 31, 53, 59, 71, 72, 159, 160
Recrystallization 19, 23, 112, 181
Refractive index 41, 82, 124, 162, 182
Regulations 142, 225, 227
Relative humidity 35, 36, 44, 45, 74, 116, 118, 147–149, 182
Reports 7, 128, 137, 138, 140, 141, 148, 195
Rochelle salt 123
Rocks 2
  Igneous 2
  Sedimentary 2

S
Samples 12, 24, 28, 42, 91, 139–145, 151–156, 160, 162, 170, 179–181, 183, 224
Saturated solution 28, 34–38, 54, 55, 68, 69, 107, 116, 149, 151
Schemes
  Flow 12, 29, 132, 170
  Notes 179–182
Seed crystals 54, 121–123, 228
Semiconductor 117
Shelf life 22, 74, 76, 202, 227
Silicates 50, 61, 219, 220
Silver halide 49, 83
Sinter 3
Sodium bicarbonate 23, 197, 217
Sodium carbonate 22, 23, 56, 72, 73, 107, 108, 193, 197, 203, 204
Sodium chloride 24, 28, 35, 38, 44, 70, 116, 134, 186, 200, 201
Sodium hexametaphosphate 65, 203
Sodium nitrate 191
Sodium phosphate 122, 204

Sodium pyrophosphate 56
Sodium trimetaphosphate 20
Solid solutions 81, 83, 97, 98, 103, 104, 107, 109–112, 152, 153, 193, 194, 200, 201, 224
Solubility 3, 5, 16, 26, 28, 33–36, 47, 55, 59, 62–82, 95, 97–122, 147–153, 192–194, 196–213, 224
Spray drying 208–215, 221, 225
Stacks 8
Static charge 115, 126–128
Statistics 75, 83, 139, 143, 144, 146, 147, 169, 225
 Experimental design 139, 145
Storage
 Bags 3, 8, 9, 22, 128, 188, 225
 Bins 6, 15, 22, 29, 223, 229
 Rail cars 8, 10, 22, 72, 137, 227
 Silos 15, 29, 186
 Warehouse 5, 8, 25, 31, 33, 121, 225
Sugar 4, 21, 44, 48, 186, 187, 199, 200, 207
Sulfates 72
Sulfur 22, 48, 84, 92, 122, 125, 166, 218
Sulfur dioxide 22, 218
Supersaturation 63
Surfactants 50, 73, 76, 107, 190, 197
Swimming Pool Chemicals 201, 202

T
Tartaric acid 120, 167
Television equipment 163
Tests 12, 28, 31, 46, 61, 62, 79, 108, 126, 142, 143, 148, 149, 165–168, 209
 Ice point 61, 158, 159
 Kundt's 121, 122, 125, 166
 Water absorption 36, 59, 110, 147

Weight loss on ignition 105, 160
x-ray analyses 161
Thermocouple 60, 154–159
Thymol 48
Tie lines 104, 105, 107, 109, 110, 152, 200, 224
Titanium salts 124, 204
TR 9
Tricalcium orthophosphate 132, 133, 186, 197, 200, 229
Triple point 159
Tripolyphosphate 9, 38, 40, 50, 56, 65, 76, 77, 85, 112, 197, 198, 204
Tunneling microscope 47

U
Urea 39, 48, 50, 110, 191–194, 198, 200, 213, 221

V
Van Wazer, J. R. 79, 80, 136, 206, 221
Vapor pressure 5, 28, 34–38, 41, 42, 48, 51, 52, 55, 110, 147, 148, 149, 159, 183, 193
Vapor pressure lowering 34, 41, 42, 48, 193
Vapor tension 28, 36, 37, 44, 45, 52, 55, 110, 147–149, 183, 193, 199
VCR 91

W
Water
 Absorbed 45–48, 55, 60, 67, 86, 107, 148, 149
 Action 10, 20, 31, 42, 43, 45, 54, 78, 88, 107, 122, 135, 158, 192, 219
 Adsorbed 43, 44, 46–49
 Atmospheric 19, 23, 29, 38, 48, 52, 60, 64, 68, 76, 93, 110, 149, 159, 189
 Constitution 160, 161

Film 28, 49, 83, 87, 182
Filters 202, 203
   Free 8, 72, 74, 75, 109, 160, 183
Generation 230
Heat capacity 25, 26, 41, 82, 156, 157
Humidity 26, 35–37, 39, 44, 45, 74, 99, 108, 116, 118, 147, 148, 149, 182, 183
Hydration 3, 5, 9, 19, 38, 41, 45, 51, 55, 56, 66, 67, 107, 151, 181, 198, 210
Hydrolysis 64, 67, 78
Leaks 10
Migration 78, 117–119
Physical properties 5, 9, 62, 73, 77, 82, 92, 138
Pure 159
Rain 10
Sink 56
Solvent 16, 34, 40, 41, 62–64, 66, 68, 69, 75, 192, 215, 224
Surface tension 41, 42, 55, 107, 133, 134, 188, 213
Treatment 18, 20, 22, 30, 38, 60, 72, 73, 84, 86, 126, 144, 145, 160, 189, 199, 203
Triple point 159
Vapor 5, 8, 9, 28, 29, 34–48, 51–55, 64, 74, 77, 78, 108, 110, 147–149, 159, 183, 189, 193, 199
Vapor tension 28, 36, 37, 44, 45, 52, 55, 110, 147–149, 183, 193, 199
Wash 73, 152
White powder syndrome 141

X
x-ray analyses 161

Z
Zero fluidity 161
Zinc nitrate 189
Zinc oxide 101

KH Eckelmann 7/94